W0193158

The Dialogue of Civilizations in the Birth of Modern Science

The Dialogue of Civilizations in the Birth of Modern Science

Arun Bala

THE DIALOGUE OF CIVILIZATIONS IN THE BIRTH OF MODERN SCIENCE
Copyright © Arun Bala, 2006.

Softcover reprint of the hardcover 1st edition 2006 978-1-4039-7468-6

All rights reserved. No part of this book may be used or reproduced in any manner whatsoever without written permission except in the case of brief quotations embodied in critical articles or reviews.

First published in 2006 by
PALGRAVE MACMILLAN™
175 Fifth Avenue, New York, N.Y. 10010 and
Houndmills, Basingstoke, Hampshire, England RG21 6XS.
Companies and representatives throughout the world.

PALGRAVE MACMILLAN is the global academic imprint of the Palgrave Macmillan division of St. Martin's Press, LLC and of Palgrave Macmillan Ltd. Macmillan® is a registered trademark in the United States, United Kingdom and other countries. Palgrave is a registered trademark in the European Union and other countries.

ISBN 978-0-230-60979-2   ISBN 978-0-230-60121-5 (eBook)
DOI 10.1057/9780230601215

Library of Congress Cataloging-in-Publication Data

Bala, Arun.
    The dialogue of civilizations in the birth of modern science/Arun Bala.
      p. cm.
    Includes bibliographical references and index.

      1. Science—History. 2. Science—Philosophy. 3. Science and civilization. I. Title.

Q125.B2925 2006
509—dc22   2006043225

A catalogue record for this book is available from the British Library.

Design by Macmillan India Ltd.

First edition: October 2006

10 9 8 7 6 5 4 3 2 1

Transferred to Digitasl Printing in 2007

Dedication

To the memory of my father and mother for their acute sensitivity
to intercultural dialogue which opened my eyes to worlds that
otherwise might have remained closed

Dedication

Contents

Preface and Acknowledgments

This book has been long in the making.

It began twenty years ago when I coordinated a project for the Southeast Asian Regional Institute of Higher Education Development, to promote environmental education throughout universities in the region. Discussions with colleagues across the region soon made it apparent to me that developing knowledge to protect the environment required opening a dialogue with the vast reservoirs of knowledge accumulated by local cultures in preserving and managing their environments. Such a view went against the grain of my training both as a physicist and philosopher of science. My disciplinary backgrounds were predisposed to view with suspicion any dialogue between science and traditional knowledge, perceiving the latter to be no more than a congeries of myth, superstition, and error.

My newly enhanced receptiveness to promoting dialogue between science and indigenous traditions informed my courses when I began teaching history and philosophy of science at the National University of Singapore. Moreover, at that time the end of the ideological Cold War gave birth to the notion of a "clash of civilizations" engendered in part by the copresence of modern and traditional cultures. This made a dialogical orientation even more important. It could prevent the notion of a clash of civilizations from becoming a self-fulfilling prophecy by showing how dialogue across civilizations could be both viable and enriching.

Nevertheless, incorporating a dialogical perspective into my teaching seemed a formidable challenge. Mainstream history and philosophy of science have been particularly reluctant to engage indigenous knowledge. Most historians of science have been profoundly Eurocentric, assuming that modern science had learned little (if anything) from non-European and traditional cultures in the past and would have even less to learn from them in the future. Philosophers of science have generally seen scientific knowledge as standing apart from traditional knowledge and as distinct from it. Although modernists and postmodernists differ in their attitude toward this distinction—modernists see science as destined to overcome and supplant traditional knowledge, while postmodernists see the two forms of knowledge as coexisting, in a kind of epistemological detente—neither sees dialogue with traditional cultures as a fruitful undertaking for science.

This orientation of the dominant schools of the history and philosophy of science made it difficult to teach the subject from a dialogical perspective. Yet I was fortunate in two respects. First, recent decades have generated a large body of literature, albeit on the academic margins, documenting the contributions of non-Western cultures to modern science, to which I could appeal. Second, my students came from a diversity of cultural backgrounds, which made them quite receptive to studying the role that intercultural dialogue may have played in the enterprise of modern science.

Progress was slow. As the course developed over the years, solid evidence supporting the dialogical perspective accumulated gradually and in bits and pieces. The situation changed dramatically when the United Nations declared the first year of the twenty-first century the "Year of Dialogue among Civilizations." This led to a flurry of activity on the World Wide Web, as participants from diverse cultural regions and backgrounds projected the contributions of their respective cultures to knowledge in general, and science in particular.

The literature on the Web was of uneven quality but, nevertheless, made it evident that the contribution of non-European cultures to modern science was far greater than hitherto suspected. Modern science was not simply enriched by such contributions—it was crucially formed by them. This more radical conception of the importance of intercultural dialogue cannot be appreciated if we continue to view the contributions of Chinese, Arabic, Indian, and other multicultural scientific ideas in isolation from one another. Only when we start to examine how they interacted together, as well as with ideas in Europe, do we begin to appreciate the crucial role they played in the birth of modern science. It is this complex, multipolar, and historically indispensable intercultural dialogue that this book sets out to document.

The dialogical history of modern science also raises profound new questions. It evokes doubts about the viability of the mainstream history of science—particularly Eurocentric intellectualist and sociocultural histories that ignore Europe's interactions with cultures beyond its borders in the early modern era. It also calls into question the tenability of modernist and postmodern philosophies of science, neither of which allows scope for the growth of scientific knowledge through dialogue. It renders problematic the position of those scientists who see receptivity to traditional knowledge as a threat to the integrity of science. It also rebuffs those ideologies that see differences in civilizations solely as a source of political conflict whereas these could in fact be a resource for intellectual progress.

Naturally a study of this nature, which developed in slow stages, accumulates a lot of debt to those who contributed to it by way of encouragement, ideas, and institutional help. I must begin by thanking the many students whose enthusiasm supported and inspired me in the initial stages of this study. Without their desire to show and question what different cultural traditions may have meant for science, and their eagerness to avoid the easy path of cultural glorification or disparagement, I would not have been able to pursue this project in its early years.

There are many others who made a more direct contribution, by reading and commenting on the various drafts as this study evolved. I would like to begin by thanking George Landow (Brown University) whose vision of the World Wide Web as opening the door to the recanonization of knowledge as a dialogical enterprise of many voices inspired my search through the Web for alternative cultural perspectives on the history of science. His reading of an early draft of this study, and recommendation to change its focus from the history of science in general to the history of the birth of modern science, made it possible for me to strengthen my arguments and tighten the dialogical case considerably.

I would also like to express my deepest gratitude to James Robert Brown, not only for inviting me to the University of Toronto but also for giving me unstinting support as reader, friend, and critic in the course of completing this project.

And my special thanks to Donald Wiebe (Trinity College, University of Toronto), without whose generous comments and help this book would have had a harder time seeing the light of day.

I also wish to record my abiding debt to friends and colleagues who read complete drafts of this work as it went through its many stages of evolution. They include Andrew Brennan (University of Western Australia), George Gheverghese Joseph (Manchester University), Lee Keekok (Lancaster University), Anjam Khursheed (National University of Singapore), and Ng Chee Yuen (Japan Foundation for Advanced Studies in International Development). I cannot overestimate the importance of their trenchant critiques, insightful comments, and personal encouragement in nurturing and shaping this study over its many stages.

There are others who helped by listening patiently and responding critically to many of the ideas developed here at countless sessions when we met along corridors, at meals, and in offices—Farid Alatas, Belal Baaquie, Zaheer Baber, Daniel Goldstick, Meng Jianwei, Cecilia Lim, K. P. Mohanan, Victor Savage, Vijay Kumar Sethi, and John Williams. I also owe thanks to Jeffery Ewener, who read parts of the manuscript and helped crucially in flexing and sharpening the stylistic presentation of this study.

I would also like to express my appreciation to organizations and institutions that gave me opportunities to explore this dialogical approach in various contexts and to present and discuss some of the key ideas in this study—the United Nations University for inviting me as a resource person for the workshop on *Science and Dialogue among Civilizations;* the Foundation for Advanced Studies in International Development in Japan for inviting me to present my views on the role of the World Wide Web as a resource for intercultural dialogue; the Institute for the History and Philosophy of Science and Technology in the University of Toronto, the Centre for the History of Science, Technology and Medicine at the University of Manchester; and the Multicultural Studies Programme at Manchester Metropolitan University.

I cannot adequately express my gratitude to the two institutions that supported me and gave me the time, opportunity, and facilities to carry this task to completion—the Department of Philosophy at the National University of Singapore, where I began this project, and the Department of Philosophy at the University of Toronto, where I completed it.

Finally, I express my deepest gratitude to Selvarani, who served as my patient sounding board over the years, and to my family, who never lost faith in this project even if they knew little about its content.

Chapter 1

Introduction

Recent years have seen the emergence of new critics of modern science who have been described, and often describe themselves, as postmodernists, multiculturalists, or cultural pluralists.[1] Their attacks on modern science emanate from a number of different academic disciplines, including ecology, medicine, history, politics, sociology, and even literary criticism. These critics reject the commonly held view that current science is a culture-neutral enterprise of universal and cosmopolitan reach and the common heritage of humanity, despite its historical genesis within the geographical confines of Europe. Instead they maintain that the knowledge offered by modern science cannot be treated as universal, but has to be seen as inescapably rooted in the cultural singularities of the West. It leads them to demand greater recognition for the large reservoirs of indigenous knowledge carried by non-Western cultures and reject as hegemonic and imperialistic the exclusive claims made for modern science.[2] Their stance is vehemently rejected by modernists, who view any accommodation with traditional knowledge as embracing archaic, mythical, superstitious, and irrational systems of thought that would return us to the medieval "Dark Ages" that preceded Enlightenment science.[3] The resultant ferocity and intensity of the debates generated has led observers to describe them as the "science wars."[4]

Yet there is a shared Eurocentric presumption that underlies, and undermines, both sides in these debates—namely, that the historical roots of modern science must lie exclusively in Europe simply because modern science developed in Europe. This is wrong. The roots of modern science are "dialogical"—that is, the result of a long-running *dialogue* between ideas that came to Europe from a wide diversity of cultures through complex historical and geographical routes.[5] More particularly we will find that the pivotal episode in the rise of modern science, the Copernican Revolution, was profoundly influenced by Arabic,[6] Chinese, Indian, and ancient Egyptian ideas and discoveries that came together in Europe over a few centuries. Such multicultural contributions not only made possible the narrow Copernican Revolution that led to acceptance of the heliocentric theory, but also the wider Copernican Revolution that led, through Galileo and Kepler, to Newton's unification of cosmological and physical theory.

Today there is wide recognition of the global interactions that created our contemporary social, political, and economic systems as Europe, through its colonizing enterprise, drew on the resources and products of a wide diversity of cultures to bind them as parts of a single world system.[7] Modern science should also be seen as a product of this same global interaction, as Europe drew on the intellectual resources of the cultures it encountered by both dominating and learning from them.[8] The voyages of geographical discovery that put Europeans in contact with many other cultures must also be seen as voyages of intellectual discovery that allowed Europeans to acquire and exploit the knowledge of these cultures. Modern science is the outcome of a dialogue within the West that drew on ideas and discoveries from many civilizations—something that neither modernist nor postmodernist conceptions of the rise of the new science acknowledge adequately.[9]

It may be feared that the originality and creativity of the West in forging modern science is diminished when it is traced to a dialogue with ideas drawn from other civilizations. But modern science was not simply the result of a passive accumulation of ideas and practices from non-Western traditions of thought. Rather, it was the outcome of a process of integrating seminal discoveries from many cultures and combining them within Europe into a new synthesis not achieved elsewhere. The Copernican Revolution became a turning point for precisely such a multicultural synthesis because its key heliocentric idea—born in ancient Europe with Aristarchus and resurrected in modern Europe by Copernicus—could only be consolidated by drawing on the philosophical, mathematical, scientific, and technological discoveries of many civilizations.

This has important implications for how we view the so-called science wars. We cannot go along with either modernists or postmodernists when they separate the West from the Rest in their discussions of the birth of modern science. We cannot agree with the way both treat modern science as something historically and cognitively alien to the traditional sciences—with zealously "antitraditional" modernists on one side, and radically "protraditional" postmodernists on the other. Contrary to what both assume, the science wars are not between traditions of science historically and intellectually disconnected from one another. If modern science has emerged through dialogue with non-Western traditions and has grown by drawing upon their resources, then we have to recognize that the Rest have crucially penetrated the West. The tradition of modern science, generally perceived as alien to non-Western traditional sciences, already incorporates a great deal of this indigenous knowledge. Since modern science has its genealogy in the histories of both the West and the Rest, we should no longer use it to sharply sunder the intellectual history of the West from the Rest.

This raises an important question: What is the significance of contemporary, postmodern, tradition-inspired critiques of modern science if they are really confronting a hybrid bred in part from traditional knowledge? The postmodern traditionalists can hardly afford to reject outright every part of modern science, since in so doing they would be rejecting elements of their own tradition.[10] Instead, these critiques are better interpreted, whatever the intent may be of those who make them, as drawing attention to valuable knowledge and insights within the traditional sciences currently ignored or rejected by modern science. They suggest that by making modern scientists open a new dialogue with traditional knowledge holders, we could fruitfully promote the growth of science in the future as dialogue did in the past.[11]

There is also a broader cultural advantage in nurturing a deeper appreciation of the dialogical roots of modern science. An understanding of the way scientific knowledge advanced through the interaction of ideas drawn from different cultures would subvert attempts to use history either to promote the hegemony of one single culture or to make the existence of diverse cultures an excuse for confrontation and conflict. Both these orientations have been promoted recently by adopting a type of historical inquiry, by no means new, that studies the past with a view to predicting the future—or rather constructs a narrative of the past that could become a self-fulfilling prophecy for the future.

One such hegemonic narrative is the work of the political theorist Francis Fukuyama, who sees the termination of the ideological conflict between socialism and liberal capitalism as heralding "the end of history." He argues that cultures that adopt modern science, with the technology and economic system associated with it, will inevitably dominate the future course of history. In turn, they will force the rest of the world to follow the same path. For Fukuyama, modern science is the logic of modernity and the directing force of history. Criticizing those who deny progressiveness in history, he writes:

> We have selected modern natural science as a possible underlying "mechanism" of directional historical change, because it is the only large scale social activity that is by consensus cumulative and therefore directional. (Fukuyama 1992, p. 80)

Thus, Fukuyama is a modernist with a vision of global civilization as ultimately driven by the inexorable force of modern science—an enterprise invented only once in Europe but impossible to resist elsewhere. He recognizes no possibility for multicultural reservoirs of knowledge to advance future science—the end of history is the end of traditional science.[12]

In contrast to Fukuyama's vision of the coming convergence of cultures, the political theorist Samuel Huntington presents a future of cultures in collision. Huntington sees the end of the Cold War as leading away from a bipolar world, in which the conflict between liberal capitalism and socialism dominated the world stage, to a multipolar world in which "the clash of civilizations" becomes paramount. According to him the clash is rooted in the radically different values that motivate civilizations—in particular their religion-inspired political, social, and moral orientations (Huntington 1996).[13]

Even though Fukuyama and Huntington do not address the concerns of the science wars directly, what they say about civilizational interactions can also be read as applying to the interaction of civilizational sciences. This links their concerns directly to those that motivate the science wars. Fukuyama's view that science drives the directionality of history, and that history will end with the triumph of liberal capitalism, parallels the commonly held opinion that modern science lies at the end of history and will displace all previous traditions of natural knowledge. It is hardly likely that Fukuyama would disagree with this; his view of science does not even entertain the notion of alternative cognitive traditions. If it did, his claim that we have come to the "end of history" by producing the way of life inevitably linked to modern science would be weaker. A new cycle of history would begin if current science gave way to another tradition of knowledge and a different way of life.

Huntington's clash of civilizations can also be seen, by extension, as a clash of civilizational sciences. According to the social scientist Ziauddin Sardar, the clash between civilizational traditions of science is the deepest dimension of the science wars, but is often ignored by those who see it only as a clash of Western traditions—either Romantic versus Enlightenment, or postmodern versus modern. For Sardar, the civilizational sciences conflict began with the dawn of modern science but has hardly been acknowledged explicitly—he terms it "the invisible science wars" (Sardar 2001, p. 120). Nevertheless, Sardar seems to suggest that this clandestine war is now becoming more visible as critics of modern science—environmentalists, feminists, multiculturalists—increasingly turn to non-Western reservoirs of traditional knowledge to overcome the limitations they perceive modern science to have.

The patterns for civilizational interaction provided by Huntington and Fukuyama do not allow for a mutually fruitful dialogue between different civilizations—and they suggest that this is also true of science. However, if it can be shown that modern Western science profited from its dialogue with non-Western traditions of science, then it would be possible to suggest that the relations between civilizations need not always be adversarial and confrontational. Civilizations could teach and learn from one another in ways that are mutually beneficial.

Thus, the answer we give to what has been called Needham's Grand Question, "Why did modern science develop in Europe and not elsewhere?"[14]—because of a dialogue with other civilizational ideas or because Europe managed to shield itself from the religious, social, cultural, and geographical factors that kept other civilizations chained to superstition, myth, and error—will crucially influence our perception of intercivilizational relations.[15] Moreover, the question itself can be divided into two parts: "Why did modern science develop in Europe?" and "Why did modern science fail to develop in civilization X?" where civilization X can stand for Chinese civilization, Arabic civilization, Indian civilization, and so on. Answering both parts of the question has often provided an opportunity for Eurocentric historians to find praiseworthy qualities within European languages, religions, and geography, or within European political, economic, and social institutions that facilitated the birth and growth of modern science, and blameworthy qualities within their non-European counterparts that obstructed it. A dialogical history of modern science would require us to reject such one-sided explanations and suggest that praise and blame need to be assigned more evenhandedly. Traditional cultures deserve praise for making discoveries that contributed to modern science, even if they are assumed to deserve blame for failing to reach modern science. Western culture deserves blame for ignoring not only discoveries in traditional cultures that became a part of science, but also traditional discoveries that continue to be ignored by modern science, even if it deserves praise for creating modern science. Let us look now look at some attempts to answer the question, Why did modern science not develop in civilization X?, without addressing directly for the moment the question of why modern science did develop in Europe.

Chapter 2

Why Did Modern Science Not Develop in Civilization X?

The question, Why did modern science not develop in civilization X? is generally entertained seriously only with regard to civilizations like the Arabic, Chinese, and Indian, which in other respects—religion, the arts, and literature and, to an extent, philosophy—are considered to have progressed and achieved as much as the West.[1] It does not occur to anyone to ask, say, Why did Christianity not develop in China, or Renaissance representational art in India? In religion and the arts we do not expect culture-neutral constructions. But since the same science is taught everywhere, and scientific knowledge is perceived as universal and cosmopolitan, it seems to make sense to ask why it failed to develop in a particular civilization. Behind the question, therefore, lies the assumption that modern science is the only science possible and could only have developed in the way it did. Thus it seems reasonable to inquire why it failed to develop in civilizations outside Europe that, in most other respects, were at least as advanced as Europe at the time of its birth. Behind the question often lies the unspoken presumption that modern science did not develop through a dialogue with non-Western traditions of science. Were it acknowledged that such dialogue contributed to both the birth and growth of modern science, and that contemporary science grew by standing on the shoulders of earlier non-European traditions, the question would seem pointless. For example, the question of why modern science did not develop in ancient Greece is rarely asked since it is already seen as building on the accomplishments of Greek science.[2]

Moreover, the question would not be asked if modern science were treated as only one local tradition of knowledge among others, without imputing to it any cosmopolitan or universal status. Then there would seem little motive to ask why Western science did not develop in China as to ask why Western art did not. It is only because modern science is considered universal that it appears pressing to inquire why it developed in the West and not elsewhere.

Linked to the notion that modern science is universal is the assumption that it has finally displaced all other traditions of science. Future scientists have no choice but to work with, and within, this tradition of science. Such a conclusion is often associated with the Eurocentric historical assumption that modern science had no contributions from non-Western cultures, so that future scientists have no choice but to become Westernized in order to practice science. We might assume that such claims are no longer explicitly advanced by scientists; that they have become receptive to the *possibility* that non-Western cultures may contain ideas that could potentially subvert modern science even if they have not done so yet. Unfortunately, such comforting complacency would be misplaced. In their influential study *Higher Superstition: The Academic Left and Its Quarrels with Science*, Gross and Levitt fall prey to Eurocentric prejudices in their otherwise cogent criticism of the extreme relativist views too often taken by postmodern multiculturalists. They write:

> Science, as the term is now understood is, moreover, uniquely associated with Western culture. It arose only once, an invention that is unlikely to be repeated in detail no matter how many other cultures and peoples eventually come to produce fine scientists. In a hundred years, the greatest theoretical physicist in the world may well be Maori or Xhosa by descent; he—or she, as may well be the case—will nonetheless be a Westerner in the most important aspect of his or her intellectual temperament. The argument may be made *pace* Steiner, that even Mozart is outdone by the polyrhythmic splendors of the Javanese gamelan or by the sonorities of Indian classical music. No such possibility exists for the multiculturist challenging the intellectual hegemony of Western science, aside from pure falsification. (Gross and Levitt 1994, pp. 219–220)[3]

Once we assume that there can be only one possible tradition of science, that this science was discovered in Europe, and that any other culture X should have been able to discover it given proper conditions—the conditions that obtained within Europe at the time it developed there—it seems reasonable to go on to inquire why science did not develop in a particular civilization X. The answer to this question has typically followed one of two routes: either an explanation in terms of a cultural lack in X or one of a cultural obstacle within X. Explanations in terms of cultural lack appeal to a series of absences in X—the lack of the Christian religion, the Greek heritage of philosophy, the Roman concept of law, a plural polity of competing states (as in medieval Europe), an incipient capitalist merchant class, and so on. The absence in other cultures of factors present in Europe is made to account for the failure of science to arise in these cultures.[4]

Explanations that appeal to the absence of facilitating factors within civilization X are often complemented by appeal to the presence of intrinsic inhibiting factors—social, cultural, and institutional—that hindered the emergence of science. Religious dogmatism, overly pragmatic philosophies lacking theoretical interests, despotic rulers, and centralized kingdoms that hindered plural discourse are all invoked to explain the failure of modern science to emerge in culture X, despite the fact that similar factors arguably prevailed within Europe at the dawn of the Scientific Revolution.[5] The effect of such studies is to interpret culture X solely in terms of absences and presences within it that failed to provide the environment to nourish scientific ideas even had they happened to emerge in incipient form.

To see how the systematic use of such an approach can distort our understanding of both traditional sciences and their contribution to modern science, let us examine closely one major study of multicultural traditions of science—*The Cambridge Illustrated History of the World's Science* by historian Colin Ronan.[6] With his specialized expertise in Chinese science, as well as his interest in the Arabic, Indian, Egyptian, and Mesopotamian civilizations, Ronan is more sympathetic than most to premodern traditions of science. He is also one of the few historians in modern times to have looked at traditions of science across many different cultures from one comprehensive point of view. To find a parallel we have to go back to al-Andalusi's *Book of the Categories of Nations,* in eleventh century Spain, which compared Egyptian, Greek, Arabic, Indian, Chinese, and other scientific traditions before the birth of modern science. A more recent study edited by Helaine Selin, the *Encyclopedia of the History of Science, Technology and Medicine in Non-Western Cultures,* is a collection of articles by many authors, and as such it does not provide a unified perspective across civilizations. Ronan's study is thus a unique contemporary attempt to look at the global traditions of science and evaluate their contributions. Yet, despite his rich and erudite accounts of non-Western traditions of science, Ronan continues to be obsessed with the presumed cultural deficiencies and obstacles that he thinks precluded the emergence of modern science in these civilizations.

Consider Ronan's historical account of Chinese science. At the end of his long discussion of the role of the state bureaucracy in science, Ronan proceeds to answer the question, Why did modern science not develop in China? as follows:

> [D]espite the fact that in so many fields Chinese science reached at a very early date a level of knowledge equal or superior to that of Europe in 1500, there was no "scientific revolution" in China: the breakthrough into the era

of powerful modern science occurred in Europe not in the East. Why should this have been? It is plainly impossible to give a categorical answer to such a question, but it may in part be connected with this same close association of science and the State bureaucracy. In China, the urgent impulse to exploration, to new discovery for its own sake never developed as it did in Renaissance Europe, and there was no aspiration to break the mould of existing orthodoxies as inspired men such as Galileo. And one of the reasons for that must lie with the prevalence of the efficient but traditional bureaucracy of China, its rules and outlook defined by Confucius many centuries before. (Ronan 1983, p.186)

Ronan points to a variety of internal causes to explain the failure of the Chinese to create modern science—the obstacles set by the state bureaucracy, the peculiar lack of curiosity in Chinese culture, the incapacity of the Chinese to value knowledge for its own sake, the conservative nature of Chinese traditions, and the obstacles placed by Confucian thought. Thus Ronan's list includes Chinese politics, psychology, society, tradition, and philosophy as impediments in the way of science. This can even be interpreted as a sweeping indictment of Chinese civilization as a whole during its period of greatest development—from the Han dynasty beginning in 206 BCE, which made Confucianism the state ideology, to the year 1500—precisely the time during which Chinese advances in science and technology continually outstripped those of Europe. Even during the Han era, which lasted until 220 CE, the Chinese made major strides in science and technology: the compass, paper, and the seismograph were all invented in this period. By 1500 the Chinese had added the mechanical clock, printing, and gunpowder to the list, to name only a few seminal discoveries. Surely these achievements cannot be the outcome of a culture profoundly antithetical to scientific achievement in the psychological, social, political, and cultural spheres.

Let us look more closely at Ronan's charge that the bureaucracy inhibited Chinese science. This fails to explain why modern science emerged in Europe precisely in those areas that developed an efficient state bureaucracy comparable in many respects to that of China—including Britain, France, and Holland during the age of the Enlightenment, when modern science was in its formative phase. Equally unconvincing is his claim that the Chinese never developed the impulse to pursue exploration and discovery for their own sake when Confucius himself, as well as neo-Confucian thinkers, repeatedly emphasized the importance of cultivating knowledge through the investigation of things. It is precisely this emphasis on the investigation of things that explains the vast strides made by Chinese science from the Han to the early modern era under Confucian bureaucracy. But Ronan's account requires us to believe that Chinese civilization eminently advanced scientific knowledge over 1,700 years in spite of an indifference to knowledge for its own sake and a Confucian aversion to innovation. Ronan's claim appears even more dubious when we consider that the Chinese not only developed a catalogue to record the progress of their inventions but also erected statues to commemorate those who made these discoveries.

Even though Ronan deprecates the capacity of Chinese culture to sustain science, he is nevertheless a largely sympathetic exponent of the history of Chinese scientific achievements. Much more censorious is the sinologist Derek Bodde. Bodde argues that Chinese science was no more than a pseudoscience, which remarkably did not obstruct technological progress despite its metaphysical, methodological, and theoretical shortcomings. Bodde blames the failure of the Chinese to develop modern science on a large catalogue of factors—the form of written Chinese, the nature of Chinese religion, the role of government and society, the structure of Chinese morals and values, and the way human–nature relations were defined in China. Although he writes that his mission is "to determine, as best as we can, what may have been the factors in Chinese civilization that either favored or hindered scientific and technological progress," his conclusions focus on obstacles rather than facilitating factors because he does not believe there was anything like a Chinese science (Bodde 1991, p. 4).[7]

It would be unfair to charge Ronan and Bodde with constructing discourses that Edward Said characterized as projecting "orientalist structures of attitude and reference" deliberately connived to define and subjugate, control and marginalize, non-European cultures and people (Said 1994).[8] They are too sympathetic to the wider Chinese cultural tradition to be suspected of such nefarious designs. Nevertheless, their drive to identify deficiencies and obstacles in a cultural tradition that both hold in high regard—precisely at the moment they attempt to account for the Chinese failure to develop modern science—demands explanation. The answer lies in their presupposition that modern science constitutes the only possible universal and cosmopolitan tradition of science. Although the Chinese were in a position to discover this tradition, given their technological advantage over Europe for millennia, they did not. It therefore seems reasonable to infer that powerful cultural factors within China must have obstructed an otherwise natural development. If we assume that the discovery of modern science is a desirable achievement—something most of us might not be prepared to deny—these factors can easily be interpreted as shortcomings of Chinese civilization.[9]

Yet, had Ronan and Bodde, following Needham, seen Chinese science as making indispensable contributions to the development of modern science, and even serving as a necessary step toward it, they would not have viewed Chinese culture so negatively. Needham took Chinese science to be a protoscience that made important contributions to modern science. His dialogical orientation led him to examine positive factors, such as the Taoist organic materialist orientation to nature, which motivated Chinese discoveries in science and technology. Moreover, observing that Chinese organic materialist metaphysics is more concordant with the discoveries of quantum physics than seventeenth-century mechanical philosophy, Needham also thought it possible that future science might come closer to the Chinese view of nature. This suggests that Needham does not take the classical tradition of science to be the final word in science. Consequently, he is not led, like Ronan and Bodde, to develop disparaging explanations for the failure of the Chinese to develop universal science.[10]

Let us now examine Ronan's assessment of the Indian tradition of science and his judgment on why Indian civilization never developed modern science. He writes:

> In the period before the Scientific Revolution, Hindu science made a num-
> ber of original contributions that were to be importantly developed in
> China, in Islam or in Europe. Nevertheless, perhaps because of the pre-
> vailing religious tone of the Indian civilization, it never developed into a
> full-fledged science, and over the past 200 years science in the Indian sub-
> continent has had a primarily Western flavor. (Ronan 1983, p. 196)

Unlike China, where the Confucian bureaucracy is faulted for the failure to achieve modern science, the blame here is attributed to Indian religious culture. In the process Ronan ignores many ideas derived from India that he himself avows, in his historical account of Indian science, as having been of seminal significance to the growth of modern science—the decimal place system with zero for numbers, trigonometric theory and methods, algebraic discoveries, surgical medical techniques, and a highly developed linguistic theory. Moreover, he also argues that the Indian atomic theories were far more sophisticated, complex, and subtle than any developed in the West until modern times. Surely it would also be reasonable to ask what positive role, if any, Indian religions could have played in these various scientific achievements rather than taking them as obstacles to the development of modern science. After all, these major achievements were made under conditions in which the Indian religions, and the philosophies associated with them, constituted dominant elements of the culture; and the decline of Indian science largely occurred in the modern period when India came under colonial rule.[11] The failure to raise the question of the positive influence of Indian religions on science, and the emphasis on religion as the cause of decline of Indian science, itself suggests a lack of balance in Ronan's cultural appraisal—an imbalance that

appears to project the modern warfare between science and religion onto classical India.

Surprising as it might appear, and in spite of his own documentation of the important contributions made by Arabic science to medieval European science, Ronan ends his discussion of this tradition by proposing another negative diagnosis:

> Yet although the early Arabs and the whole Islamic world studied science and made notable contributions, their achievements came to an end; they never extended to modern science. Islam extols the value of revelation above all else: it is the supreme authority ... There then developed the attitude of passive acceptance. This attitude was inevitably inimical to independent scientific thinking, as intellectual traditionalism won the day. Islam never separated religion and science into watertight compartments as we do now, and the torch of science had to be carried on by others. (Ronan 1983, p. 240)

There is no doubt that Arabic science did decline after a period of brilliant achievements. Many Arabic scholars and historians had assumed until quite recently that the glorious era of Arabic science lasted about five centuries—from about 750 CE to 1258 CE—until the Mongols destroyed Baghdad and put an end to the Abbasid Caliphate.[12] However, there have been some attempts to revise such histories—it is now known that many seminal discoveries in mathematical astronomy were made in Iran in the thirteenth century by the Maragha School associated with the names of al-Tusi and al-Urdi, and that this astronomy reached its culmination in the work of al-Shatir in Damascus in the fourteenth century. Nevertheless, it is important to note, even if we ignore these later developments, that we are here concerned with a long period of scientific development—indeed a period longer than that which separates us from the publication of Copernicus' seminal *De Revolutionibus Orbium Coelestium* in 1543. Over this extended epoch it was the religion of Islam that provided the umbrella under which the sciences and philosophy developed. It inspired Arabic scholars to be receptive to scientific traditions from all the different cultures with which they came in contact—the Greek, the Indian, the Persian, and the Chinese.[13] During this period, Islam functioned as a facilitating rather than an inhibitory factor on scientific progress. Hence, to see the religious inspiration that promoted and nourished Arabic science purely as an inhibitory factor is surely to take an excessively one-sided view of the influence of Islam on science. There is, however, no denying that the mystical turn in Islam that began to dominate Arabic culture after the collapse of the Abbasids did view scientific and philosophical thought with suspicion, but this cannot lead us to ignore the positive influence of religion over a period longer than the historical era of modern science. Moreover, as we will find later, these new theological interpretations that undermined the Arabic tradition of science nevertheless paved the way for the rise of modern science in Europe (largely by undermining the metaphysical foundations of Greek and Arabic rationalist approaches to natural philosophy).

In his account Ronan seems to be adopting what can be described as an inverse-Merton thesis.[14] The sociologist Robert Merton had argued in the 1930s that Puritan religious values had played a positive role in the development of early modern science. Although his thesis was originally confined to the context of Britain, it soon became extended more widely by his followers to account for the rise of science in Europe. In contrast to these neo-Mertonian attempts to explain the rise of science in Europe by invoking the presence of a certain complex of cultural and religious values, Ronan appears concerned with explaining the failure of science to emerge in other cultures by appeal to their distinctive cultural and religious values. In particular he invokes the negative impact that Confucian, Hindu, and Islamic values played in inhibiting the emergence of modern science in cultures otherwise as advanced, or more advanced, than Europe. The Mertonian positive valuation of the Puritan religious orientation is now transformed into a thesis regarding the inhibitory role of Confucian, Hindu, and Islamic ideals on scientific growth. Both approaches are linked by their common appeal to ethical or religious values, with one difference: Merton appeals to such values to explain why modern science emerged and grew in certain places in Europe, whereas Ronan appeals to them to explain why modern science did not emerge in other places.[15]

A distinctive one-sidedness moves Ronan's overall approach. Although he acknowledges that there was scientific development in China, India, and the Arabic worlds when they were dominated by Confucian, Hindu, and Islamic ideals—indeed this is what his historical account of the sciences in these cultural areas describes—he nevertheless fails to consider the positive role these ideals may have played in the scientific achievements made there. Instead he concentrates on how these religious cultures obstructed the emergence of modern science. Thus, instead of looking at the way religious values promoted the emergence of the different structures of science in the Indian, Chinese, and Arabic cultural arenas—scientific traditions that he himself acknowledges made important contributions to modern science—he appears more concerned with how they inhibited modern science from developing there. Such an orientation can only be natural to someone who assumes that modern science is the only kind of universal science possible and that it would have developed in any culture provided there were no specific inhibitory factors precluding it.[16]

Moreover, if Ronan had taken more seriously their contributions to modern science rather than their failure to reach it, he might have been led to a more positive valuation of their norms and institutions that made such contributions possible. This becomes evident when we consider his treatment of Greek cultural values and institutions. In this instance he is led to pay glowing tribute to the Greeks, despite their failure to create modern science, because of the seminal contributions they made to it. His focus here is not on why the Greeks failed to create modern science but on why they were able to make important contributions to it. If not for the different approach he takes, and the ensuing evaluation he makes of the role Greek science played in modern science—an emphasis on their success in contributing to it—he would have been led, given his general approach to the other traditions of science, into another attempt to explain the failure of modern science to emerge in the Greek world. This might even have ended with a listing of Hellenistic cultural impediments and deficits that led to this failure. However, given his positive valuation of the Greek role in modern science, this is not the route Ronan takes. Instead, he writes:

> With Ptolemy we come to the last of the great figures of Hellenistic science, and the end of an astonishing intellectual development that was Greek in origin. It started, as we have seen, with the philosophers who wished to make sense of the physical world in which they found themselves. They would not of course have called themselves scientists—the word is a nineteenth-century one—though they might have accepted the seventeenth-century name natural philosopher, yet it was science that they practiced; not the mathematically oriented experimental science that we have today but science nevertheless, an attempt to rationalize the world of natural experience without recourse to divine intervention. This we have traced from Thales to its high development in Athens at the Lyceum and at the library and the museum in Alexandria. It marks the first great concerted effort of Western man to understand the workings of Nature … Today we can look back and see in it the foundations of the scientific ideal—the pursuit of science untrammeled by political or religious restriction—and the basis of our present scientific culture. (Ronan 1983, pp. 123–124)

Ronan's account of Greek science merits closer examination. First, although he admits that Greek science is not the mathematical empirical science we have today, he is nevertheless prepared to concede to the Greeks the achievement of having developed the foundations of modern science. Apparently, when it comes to the Greeks, science need not necessarily be reduced to modern science. The reason for this assessment seems to be that the Greeks pursued knowledge without confronting any political or religious restrictions and provided the basis for the later emergence of modern science.

However, Ronan still leaves unexplained the failure of the Greeks to achieve modern science. There is scope here for a negative diagnosis of Greek culture that he conspicuously fails to take up. Instead he stresses the values and institutions that lead to the successful growth of Greek science—precisely the sort of explanation that he does not offer to account for the growth of Chinese, Arabic, or Indian sciences in their own phases of progressive development. The fact that Greek science, although different from modern science, contributed significantly to modern science is seen as grounds for a positive judgment of Greek cultural values—but the same courtesy is not extended to other traditions, although he acknowledges that they did contribute to modern science.

Second, although Ronan acknowledges that the Greeks were more philosophers than scientists, he is prepared to call them scientists because their effort constituted "an attempt to rationalize the world of natural experience without recourse to divine intervention" (Ronan 1983, p. 123). If this is a reason for calling Greek natural philosophers scientists, then surely he should accord thinkers within Chinese civilization—especially Confucian thinkers—the same status. They, like the Greeks, developed an organic conception of the cosmos; like Plato and Aristotle, they also explained phenomena in terms of five elements—earth, water, fire, wood, metal—in contrast to the Greek elements of earth, water, fire, air, and *quintessence;* like the Greeks, their medical traditions combined a five-element theory with the qualities hot, cold, wet, and dry.

Moreover, the Chinese even went further in the direction of nonteleological explanations, because unlike Plato, who believed the heavenly bodies to be Gods, or Aristotle, who perceived them as moved by *intelligences,* the Chinese saw them in completely naturalistic terms. Nor did they separate heavenly and terrestrial phenomena by treating them as being obedient to different laws in the way the Greeks did; or take them to be made of a different sort of substance.[17] The same elements combined in the heavens as on earth.[18] This contrasts sharply with the Greek view that phenomena on earth were to be explained by the four elements earth, air, fire, and water, but those in the heavens were to be explained by the peculiar properties of the celestial element quintessence.[19] Surely if we are to use the criterion of "rationalizing the world of natural experience without recourse to divine intervention," then the Chinese have a greater claim to being scientific than the Greeks. By contrast Plato, with his heaven filled with Gods, and Aristotle, with his notion of divine *intelligences* guiding heavenly motions, appear only as quasi-scientific thinkers.[20]

Why then does Ronan see Confucianism as an inhibiting factor that prevented the Chinese from developing modern science but nevertheless fail to perceive the quasi-animist views of Plato and Aristotle similarly?

The answer seems to be indicated in the very passage quoted earlier—Greek science contributed to the birth of modern science. Hence the Greeks, in spite of having an order of science different from that of modern science, did possess those positive cultural values that promoted the birth of modern science. By contrast the Chinese—whose science is well known to Ronan since he specialized in the study of it—did not contribute to modern science in the same deep sense. It behooves us therefore, so Ronan seems to think, to examine the cultural factors that precluded the development of modern science in China. Such an account fails to value Chinese science positively because of its failure to make crucial contributions to modern science, but it values Greek science precisely because it did. It is this ultimately Eurocentric bias in evaluating their contributions that leads Ronan to see Greek cultural values as positive and Confucian ones as negative with regard to science.

Moreover, Ronan's Eurocentric bias leads him not only to praise the Greeks for their contributions to modern science, and disparage trans-Western cultural values for obstructing the emergence of modern science, but also to dispute the claims by many ancient Greeks of their indebtedness to scientific discoveries rooted in Mesopotamia and Egypt—especially the latter. He counters the prevalent Greek view that their mathematics began in Egypt:

> The Greeks claim to have received their mathematics from Egypt by way of Thales; Herodotus, Aristotle and his pupil Eudemos, who wrote a history of mathematics, all claim that Thales "after a visit to Egypt, brought this study to Greece." Indeed, Eudemos goes so far as to specify what Thales brought, namely a number of propositions in theoretical geometry. Yet our present knowledge of Egyptian mathematics gives us no grounds for supposing that the Egyptians actually possessed any geometrical theory; theirs was a practical rule-of-thumb geometry. But if Thales did not bring such geometry from Egypt did he, perhaps, devise it himself? Certainly geometry was that branch of mathematics at which the Greeks were to excel: their art shows the love of symmetry and elegant shapes but, though they were later to display geometrical genius, there is no firm evidence that Thales began the process. Certainly he did some geometry, but it would all seem to have been of a practical kind—and that was just the kind of thing he could have brought from Egypt. Yet, of course, it was from such practical geometry that the whole theoretical structure was later to be developed. (Ronan 1983, p. 68)

Ronan's interpretation is extremely curious. He dismisses the Greek accounts of their debt to Egypt and yet gives no grounds for rejecting the authority of Herodotus and Aristotle except to say that their views cannot be substantiated on independent grounds. One is led to wonder what other grounds are needed. These thinkers are closer to the events being described than Ronan, who has no hesitation in claiming that the Egyptians could not have made the contributions they were said to have made. Instead Ronan offers the weak argument that the Greek talent for geometry came from their love of symmetry and elegant shapes in their art. On such grounds the Egyptians should display similar talent—their art and their monumental pyramid architectures display an equal love of symmetry and elegance. His whole exercise of arbitrary dismissal without presenting any counterevidence to claims by dependable Greek writers seems solely designed to support his opinion that theoretical discourse in geometry first emerged with the Greeks, thereby preserving his Eurocentric conception of the history of modern science.

There is a more charitable way of interpreting Ronan's claim. He could be saying that Thales did not bring geometry from the Egyptians in the sense of a geometrical theory with associated theorems and proofs, but only what Eudemos calls a number of "propositions in theoretical geometry"—parts of mere "practical rule-of-thumb geometry." Why should this be grounds for saying that the Egyptians did not possess geometry or mathematics? They can easily be seen as the first seeds of theoretical geometry that became much more systematized by the Greeks. One cannot deny that the Greek achievement of casting geometry as a deductive system provided a richer and more rigorous theoretical framework for the discipline, but could this discipline even have taken off without building on the creative discoveries of the Egyptians that Ronan dismisses as mere practical rules of thumb? Indeed one could even argue that discovering an interesting theorem in geometry is far more difficult than proving it—the former involves a visionary experience that cannot be predicted in advance; the latter, a much more directed activity whose end goal is already predetermined. Even in the modern era the mathematician Ramanujan provides a classic example of one who made numerous discoveries in number theory that many lesser mathematicians spent decades proving. If Ramanujan's achievement can be deemed mathematical, there is no reason not to extend the same courtesy to the ancient Egyptian thinkers who came up with the theorems of geometry that Thales carried to Greece.

The logic of his position—namely that only Greece achieved theoretical knowledge in mathematics and science even when it learned something from others—also leads Ronan to write off the significant contributions that Mesopotamian civilization made to the development of scientific thought. Thus, after writing admiringly "that the Sumerians and Babylonians made great advances in mathematics may be an understatement; they seem to have laid the very foundations of the entire subject" (Ronan 1983, p. 41), he forgets his high praise when he finally comes to evaluate their achievements:

> Yet for the most part the activities of these ancient astronomers and mathematicians eventually proved to be misdirected. Their work in observing and calculating the movements of the heavens was often astonishing, but much of it proved of little importance to the ultimate development of science. It was their failure to inquire more carefully into the nature of the heavenly bodies, and into the nature of the mechanisms that drove them across the sky, that led to this dead end. Nevertheless, the subtlety of their work is a fine tribute to the inspiration of animistic religion, and of the belief in the kind of magic that ascribes more importance to the relationships between phenomena than to the nature of the objects themselves. (Ronan 1983, p. 61)

Ronan's account of the history of science appears seriously muddled on this point. How can the achievements of the Sumerians and Babylonians, who, in his own words, "laid the very foundations" of mathematics, turn out, in his own words, to be of "little importance to the ultimate development of science"? There is a profound gap between the high praise given earlier in the text and the disparaging conclusion that follows. This conclusion appears solely designed to erect walls between the Greek and earlier traditions of science so as to preserve the Eurocentric history of modern science that informs his whole, otherwise rich, study. To admit that Sumerians and Babylonians could have laid the basis for mathematics is to open the door to a dialogical conception of science subversive of his Eurocentric paradigm of its history.[21]

Ronan's history of Chinese, Arabic, Indian, Egyptian, Mesopotamian, and other traditions of science concentrates too much on faithfully recording their achievements while at the same time marginalizing any significant appreciation of their impact on modern science. Non-European traditions of systematized knowledge of nature are acknowledged; but their influence, even when noted, is minimized. Ronan seems bent on showing that modern science developed in Europe because cultural obstacles precluded it from developing elsewhere; and that though it received contributions from many cultures, only Greek culture carried those positive cultural values that made modern science possible.

Consider Needham's Grand Question "Why did science develop in Europe and not elsewhere?" in the light of Ronan's approach to the history of science across civilizations. His Eurocentric response to this question recognizes the role of dialogue with Greek science that led to the birth and growth of modern science, but marginalizes the influence of other cultures as not crucial, even when he acknowledges them. It leads him to look for positive values in Europe that contributed to the rise of modern science and negative values in other cultures that obstructed it. Such Mertonian and inverse-Mertonian explanations of the rise of modern science, and its failure to emerge, can only be subverted by demonstrating the *indispensable* role played by ideas from many cultures in the birth of modern science—especially in the success of the Copernican Revolution, which can be deemed to be the key event that led to modern science. This is precisely what we shall set out to accomplish in the present study. However, before we undertake this task, it is illuminating to examine more closely the historical and epistemological presumptions that have informed many otherwise different Eurocentric histories of modern science.

Chapter 3

The Eurocentric History of Science

By a Eurocentric history of science I mean any account of the birth and growth of modern science that appeals solely to intellectual, social, and cultural influences, causes, and ideas within Europe, and that marginalizes the importance of contributions, if any, of cultures beyond Europe to the birth and growth of modern science. Indeed, until quite recently, the possibility that Europe could have been crucially influenced by other cultures in constructing modern science was hardly entertained.[1] A typical view of this kind is expressed by Rupert Hall:

> Europe took nothing from the East without which modern science could not have been created; on the other hand, what it borrowed was valuable only because it was incorporated in the European intellectual tradition. And this, of course, was founded in Greece. (Hall 1962, p. 6)

The situation began to change with the publication of Needham's monumental series *Science and Civilization in China.* For the first time Needham and his collaborators presented an impressive body of evidence of both technologies and ideas that had originated in China and, after transmission to the West, crucially conditioned the birth and growth of scientific ideas there in the modern era. Though Needham's view regarding the extent of the influence has been contested, he opened the door to the possibility that the development of science within Europe may not have been as insular a process as hitherto assumed. There have been since then widening attempts to document the influence of non-European cultures on the development of modern science. Nevertheless, these dialogical counterperspectives remain on the periphery, and dominant histories of modern science continue to remain Eurocentric.

Although the Eurocentric position is orthodox and well established, any attempt to identify it more precisely immediately leads us into a host of difficulties. What appears to be a clearly defined position begins to lose its sharp definition as we attempt to make a closer approach. Even among those who share the Eurocentric presumption that the history of modern science is the outcome of ideas, values, institutions, and practices autogenerated within Europe, there exists considerable divergence in views about what these actually are. One major divide is between internalists and externalists. Internalist historians explain the changes that led to modern science by invoking intellectual factors such as philosophical, methodological, and scientific ideas. Externalists seek for causes in the social, political, and economic conditions within Europe at the time of the birth of modern science.

Even inside these broad categories, there are further divisions. Among internalists, some stress the role of method in shaping the growth of scientific knowledge; others emphasize metaphysics; and some see the significant variable as a scientific theory. Historians like Whewell lay stress on the inductive method—what he terms the explication of conceptions, the colligation of facts, and the consilience of inductions. Kant emphasizes the importance of the change brought about by Bacon's active experimental method in contrast to the passive collection of facts that had been the guiding strategy of Aristotelian science.[2] In the early years of the twentieth century, historians such as Koyré (1957), Burtt (1959), and Dijksterhuis (1961) believed that the emergence of science can be attributed to a metaphysical orientation that promoted the mathematization of nature. They identified Galileo as the seminal figure who promoted this turn by arguing that mathematics constituted the language of nature. Others such as Duhem (1985) and Kuhn (1957) go beyond such metascientific explanations and see the rise of science in the development of a specific scientific theory. For Duhem the impetus theory in the fourteenth century, and for Kuhn the Copernican theory in the seventeenth century, played the key role in the Scientific Revolution. However, even if they do not see metaphysics and method as major factors, they do not ignore them altogether.

In contrast to the internalists, who hold positions that at least can be categorized into clear-cut alternatives even if these sharp distinctions become fuzzy as we examine their positions more closely, the externalist views are so diverse and multifaceted that they elude categorization altogether. Religion, technology, history, geography, politics, economics, society—and in the not-too-distant past, even race—have been invoked as factors contributing to the emergence of science from conditions within Europe. Hooykaas appeals to the biblical worldview, Merton to Puritan values, Marxists to the needs of an emerging capitalism, Landes to the revolution in the conception of time brought about by the clock, and so on. Going through these diverse explanations one gets the sense that nearly every factor conceivable has been identified as crucial for the emergence of modern science in Europe.[3]

More recently some historians of science have even contested the notion that there was a scientific revolution within Europe that led to the modern era. They argue that there is so much continuity in the science that followed Copernicus with the science that preceded him that it is difficult to conceive of any revolutionary break with tradition separating them. One of the pioneers of this approach is Herbert Butterfield who, in his seminal study *The Origins of Modern Science*, traces the beginning of contemporary science to the thirteenth century. The social constructivist Steven Shapin also claims that his book *The Scientific Revolution* was really about showing that there was no such thing.[4] Peter Dear, in *Discipline and Experience: The Mathematical Way in the Scientific Revolution*, takes a similar stand. Dear acknowledges that many mathematicians and natural philosophers, such as Mersenne, Descartes, Pascal, Barrow, Newton, and Boyle, did break with the heritage of the medieval period, but it had less to do with the contents of their claims and more to do with their appeal to personal experience than to traditional authority. However, the changes in ideas they wrought did not occur suddenly, or break radically with the past. Rather, the shift of ideas involved a much more gradual and halting process that extended over more than a century. Hence Dear rejects the notion of a revolutionary intellectual mutation leading to modern science. These new histories lead Margaret Osler (2000), in the introduction of her edited work *Rethinking the Scientific Revolution*, to wonder whether the notion of a scientific revolution may not itself be a construct of historiography with its narrative of canonical heroes, such as Copernicus, Kepler, Galileo, Newton, and canonical subjects, such as astronomy, physics, and mathematics.[5]

These no-revolution views have been attacked by the social historian Howard Margolis in his study *It Started with Copernicus: How Turning the World Inside Out Led to the Scientific Revolution*. He maintains that the notion that there was no such thing as a scientific revolution does not make sense, especially if we note the fact that within Europe there was barely any noticeable change for fourteen centuries before most of the major discoveries associated with the rise of modern science came to be made within a few years around 1600 CE. This change was nothing short of revolutionary and requires explanation—it cannot simply be explained away. He attributes the shift to a novel transformation in the psychological orientation of the canonical heroes of the revolution—every one of whom came to be inspired by Copernicus' idea of a sun-centered universe. The counterintuitive claims of the theory led them to look, as Aristotle and the ancient Greeks did not, for hidden evidence and explanations for phenomena—even though such evidence and explanations had always been available at hand but had gone unnoticed. In this respect, the revolution could have happened earlier but did not—until Copernicus turned the world inside out and triggered a new orientation to nature.

Yet what is striking about the diverse internalist, externalist, and no-revolution approaches is their silence concerning the impact of non-European cultures on the Scientific Revolution in Europe. This even includes recent theorists such as Shapin, Dear, and Osler, who question the notion of a scientific revolution, as well as their critics such as Margolis—all of them ignore much of the recent literature documenting multicultural impacts. Even where they are prepared to acknowledge that some ideas did enter Europe from China, the Arabic world, or India, these are not seen as being crucial. Their accounts easily lead us to suppose that the transition to modern science from medieval science could have occurred without such multicultural contributions, even though European thinkers exploited them opportunistically when they were available. Thus the new historians also assume, like their predecessors of earlier decades, that modern science is not the outcome of vital contributions of ideas and influences from cultures outside Europe.

Indeed there is often great resistance to probing more carefully the role of such influences even when their possibility is acknowledged. In his study *The Scientific Revolution: A Historiographical Inquiry,* historian Floris Cohen acknowledges the possibility of dialogical influences but quickly dismisses it by questioning the motives of those who raise the issue. Approaching Needham's Grand Question of why modern science emerged in Europe, he writes:

> One way out has been to deny that early modern science was a uniquely Western accomplishment and to argue that its key concepts may already be discerned in ideas originally developed in Islam or in China.
>
> A more fruitful and realistic way out has been to develop standards for a truly *comparative* history. A history, that is, that takes the fact of Western priority in this particular domain of human achievement as a fact, not one to be unduly proud (or envious) of, but just as a remarkable fact that cries out for scholarly explanation through finding out how it is that the Scientific Revolution eluded other civilizations. If approached along such lines, the question is meant to throw light, not only upon Western civilization by bringing into sharper relief how it could produce early modern science, but also upon civilization p or q by focusing on features uniquely its own less conducive overall to science in the post-17th-century sense. (Cohen 1994, p. 404) [Cohen's emphasis]

What is noteworthy about this passage is that Cohen recognizes the possibility of a dialogical perspective on modern science that explains its birth in Europe through a convergence of key ideas from cultures outside. But rather than pursuing this line of investigation further, he dismisses it abruptly. In place of such a dialogical approach that might examine the exchanges with Chinese and Arabic civilizations that could possibly have led to modern science in Europe, he recommends a comparative history. He advises us that the only fruitful approach is to take "the fact of Western priority in this domain of human achievement as a fact ... a remarkable fact"—and compare those factors within Europe that promoted the rise of modern science with those outside that precluded it. In short he advises us to take the Eurocentric model as a fact without even considering the possibility of a multicultural dialogical alternative.

Of course, it can be argued that Cohen's emphasis on the so-called fact of Western priority need not necessarily imply that this "fact" precluded influences from other cultures. However, a closer reading of his study reveals that this is indeed the goal of his historical construction—he means to show that modern science developed within Europe with only marginal influences from other cultures. Moreover, he recommends in the passage quoted above that we should not be unduly proud in accepting, or unduly envious in rejecting, the "fact" of Western priority.

Surely if there is the possibility of a cross-cultural influence, then it merits scrutiny without being dismissed out of hand as motivated by envy triggered by Western pride in the fact of priority! Moreover, the "Western priority" that Cohen is intent on preserving would by no means be diminished or subverted by the discovery that ideas from other civilizations were incorporated in its synthesis of modern science—no more than the priority of the creators of modern science is diminished by Cohen's view that they were inspired by ideas from Greek science. On the contrary, making modern science have roots in traditions of knowledge beyond Europe would give it more global significance. A culturally hybrid history of modern science would make it much harder for other civilizations to reject it as a completely alien implant from the West.

By adopting the Eurocentric model on principle, Cohen is forced by his strategy to do what he wants to avoid—find praiseworthy features in Western civilization that led it to produce modern science and blameworthy factors in others that made them less conducive to science. Had he asked the dialogical question, What led Europe to discover modern science by creatively deploying ideas and practices from other civilizations? he would have been led to look for positive factors within Europe that made it receptive to ideas from the outside, and enabled it to transform and assimilate them so as to create modern science. He would also have been led to wonder what positive factors enabled the influencing civilizations to create those ideas and practices that subsequently turned out to be so fruitful for modern science. Such a historical approach would disparage neither European nor non-European contributions to modern science; furthermore, it would find reasons to praise both in different ways.

Of course we cannot simply assume in advance that there were indeed crucial influences from the outside that shaped the growth of modern science in Europe any more than we can allow Cohen to assume that there were no such decisive influences—we would need to demonstrate it. But an inquiry along such lines cannot even begin if we follow Cohen and reject the possibility of multicultural influences in advance. If other cultures did contribute in significant ways to Europe's synthesis of modern science, then one cannot follow Cohen and begin the historical investigation of the rise of modern science by taking it as a fact that they did not, and confine ourselves to comparative history. We need to examine the ways in which ideas across cultures interacted to produce modern science. If it is the case that there were significant multicultural contributions to the synthesis of modern science, then acknowledging them would neither be motivated by multicultural envy nor diminish European pride, as Cohen seems to suggest in the above passage. Envy would be an issue only if the multicultural influence had not occurred; pride would be affected only if it were claimed that Europe did not create modern science. A dialogical account acknowledges both the multicultural roots of modern science and its achievement as a creation by Europeans within Europe.

Nevertheless, before looking into the dialogical perspective, it would be illuminating to examine more closely Cohen's Eurocentric account. Cohen himself develops his position after undertaking a thorough and comprehensive study of various attempts to understand the birth of modern science—in short, after a wide-ranging survey of the history of histories of modern science. Although he punctuates his study with disclaimers that his historiography does not theorize the positions of the thinkers he examines—including Kant, Whewell (1847), Duhem, Koyré, Dijksterhuis, Burtt, Yates, Butterfield, the Halls, Kuhn, and Westfall—it is crucially directed by Needham's Grand Question, Why did modern science develop in Europe and not elsewhere? However, by precluding in advance any significant multicultural influences in shaping the rise of modern science, Cohen is led to answer the question by reference only to cultural values and institutions within Europe. This is precisely the sort of Eurocentric answer we have seen offered by Ronan. Nevertheless, Cohen's careful scholarship and faithful concern to present with integrity views that he obviously disagrees with makes his account—despite its Eurocentrism—both informed and nuanced.

Cohen begins by qualifying his historical account as a highly provisional sketch. He then proceeds to describe how ideas about natural phenomena first emerged within traditional societies in the course of attempts to make sense of the world. According to him it was by going beyond this early mythical stage that two civilizations—China and Greece—developed a systematic body of ideas to explain natural events. Following Needham, he argues that the Chinese developed the natural philosophy of organic materialism—a framework capable of yielding important insights about the world but that, Cohen maintains, ultimately turned out to be a dead end. By contrast the Greeks were able to transform loose computing rules in mathematics and astronomy, discovered by Egyptian and Mesopotamian cultures, into an abstract system of deductive geometry and to articulate a wide spectrum of broad conceptions about the cosmos culminating in the Aristotelian synthesis—a framework able to explain and understand a vast body of empirically collected information. Although Cohen acknowledges that mathematics did not play a major role in the Aristotelian corpus, he argues that this lack was compensated for by the Platonic stream of Greek thought. Platonic geometrical ideas inspired a drawn-out movement extending over centuries to mathematize—more precisely, geometrize—the disciplines of statics, optics, and astronomy, which after their climax with Ptolemy saw the growth of Greek science come to an end.

Cohen claims that three civilizations inherited this Greco-Hellenic tradition—the Byzantine, the Arabic, and the Western European (albeit the last civilization inherited this only through Arabic intermediaries in a later era). The Byzantines contributed little to enrich the tradition they inherited. The Arabic civilization added to it the decimal place system (taken from the Indians), improved some of the parameters in mathema- -tical astronomy, and raised speculations concerning local motion, but remained confined to the Aristotelian framework that it took for granted. However, Arabic science did take further the Archimedean tradition of solid and fluid statics. With these accomplishments Arabic science came to an end.

In the meantime, after the collapse of the Western Roman Empire, another civilization arose in Western Europe that was to transform the Greco-Hellenic tradition. Although it had originally inherited only a small portion of Hellenic science, it vastly expanded this patrimony by transla- tions of Arabic and Greek works in the twelfth and thirteenth centuries. Since its original heritage was largely Platonic and rationalist, it now selected the Aristotelian, empirically oriented synthesis to satisfy its new intellectual needs. Within the Aristotelian framework some problems like free fall and projectile motion were addressed—later to become central issues whose resolution led to the subversion of ancient Greek science— but they remained marginal questions designed only to fill the perceived gaps, both rational and empirical, within the Aristotelian framework.

With the fall of Constantinople to the Turks in 1453, and flight of Byzantine scholars westward, there emerged another opportunity for Europeans to appropriate the whole of the Greek corpus—especially the Archimedean tradition with its Platonic mathematical orientation. The works of Archimedes were translated in 1543—the same year in which Copernicus published his heliocentric planetary theory. According to Cohen these works had a profound impact because they encouraged some thinkers to put into question the Aristotelian orientation to nature, which had dominated European thought since the thirteenth century.

By contrast the Copernican theory did not make an immediate impact. Although it did provide a simpler mathematical model than the geocentric theory, it was generally perceived to be physically implausible. It was unable to answer simple questions like Why do bodies fall to the earth? On the geocentric account there appeared to be a reasonable answer: heavy bodies were drawn naturally to the center of the universe located at the center of the earth. Nor could it answer questions like why an earth spinning on its axis at such speed that it completed its revolution every twenty-four hours did not give rise to raging storms on the oceans or torrential winds. The absence of stellar parallax in spite of the vast distances covered by the earth when it revolved around the sun was another serious objection, since it was assumed that the stars were nearer than we know them to be today. These unanswered questions led many astronomers to treat the Copernican theory only as a mathematical hypothesis able to "save the phenomena"—not as a physical theory that described how the heavens moved.

Cohen goes on to argue that the situation changed when Galileo and Kepler—under the inspiration of the Archimedean mathematical approach—overcame all obstacles to defend the Copernican theory on realist grounds. Kepler, with the indispensable help of Brahe's more accurate and systematic astronomical data, created a new model of planetary motion based on ellipses that did away with the fictional circles used by Ptolemy and Copernicus—circles designed to ensure conformity to Plato's dogma that all heavenly motions should be circular.[6] Galileo's genius lay in his resolution of the problems of free fall and projectile motion rooted in Aristotelian physics that served as objections to Copernican astronomy. Galileo turned what appeared to be objections into a virtue by deploying Archimedean techniques that appealed to ideal situations only mathematically conceivable—such as perfectly smooth planes along which objects moved—to develop the basis for a new physics compatible with a realist interpretation of the Copernican theory. Between them, Galileo and Kepler ushered in what Cohen calls a wholly new "form of life"—the practice of science within a mathematical universe of precision in contrast to the universe of "more-or-less" that characterized Aristotelian physics.

However, Cohen also argues that the new universe of precision would not have been able to sustain itself, let alone expand as it subsequently did, were it not the case that there was already in place in Europe a set of attitudes that favored its reception—a complex he refers to as "the European Coloring" (Cohen 1994, p. 509). First, there was the urge to make very accurate observations of natural phenomena. This expressed itself in disparate areas, such as geographical, botanical, and anatomical descriptions, and also in Brahe's penchant for careful astronomical observation that discovered the data which made Kepler's achievements possible. Second, there was the tradition of the application of mathematics to art by Renaissance artists. The universe of precision fitted in well with this mathematical aesthetic orientation. Third, the positive valuation of manual labor in Europe favored the reception of the new way of life—it allowed Europeans to be receptive to new empirical technologies and alchemical practices and to treat them as a resource for theoretical investigations. Fourth, in Renaissance naturalism there was a greater receptiveness to magical notions and practices. This fourth factor is linked to the last element of the European Coloring Cohen identifies—the rise of iatrochemistry, which transformed the enterprise of alchemy from that of the pursuit of gold to the pursuit of health. The last two elements—the receptivity to magical notions and the rise of iatrochemistry—are not, Cohen maintains, directly linked to facilitating the reception of the universe of precision in Europe. However, they led to the Baconian experimental sciences that interacted with the mathematical realist orientation of the universe of precision in ways that proved extremely fertile for the future development of modern science.

These dramatic changes led European thinkers to seek for a new philosophy of nature to replace the collapsing Aristotelian one. According to Cohen the resulting endeavors focused on two possibilities. One was the corpuscular philosophy ultimately derived from the Greeks; the other was the Hermetic tradition and chemical views associated with the European Coloring. Ultimately, corpuscularism was to triumph because it was found to be more consonant with the mathematical universe of precision—though not without some expenditure of effort, since it did not involve the mere appropriation of Democritean atomism.

The combination of the concept of the universe of precision with the European Coloring led to two novel thought constructions. First, it inspired Bacon's ideology of science with its marked utilitarianism rooted in the Hermetic conception of man as a magician commanding the powers of nature. Second, it led to the "Baconian" experimental sciences concerned with studying nature by adopting the method of active experimentation rather than the passive collection of data. The experimental approach was especially crucial in areas concerned with the study of chemical reactions, magnetism, and static electricity. However, even in these areas the influence of the universe of precision is apparent. Cohen writes:

> As the [seventeenth] century moved on, these two latest products of the European Coloring—the Baconian sciences and the Baconian ideology—were rapidly shorn of the "spiritual" atmosphere out of which they had arisen in the first place; here again the influence of the universe of precision made itself felt. (Cohen 1994, p. 513)

Over time, he continues, the Baconian complex—the sciences plus the ideology—merged with the corpuscular philosophy. This happened around the middle of the seventeenth century with the process culminating in Boyle and Hooke, who combined heuristic experimentation with the corpuscular conception of nature. It was from this brew (and particularly its alchemical component) that Newton distilled his novel conception of force as an active principle. By subjecting this concept to experimental testing and mathematical rigor, he united the universe of precision with the Baconian tradition. It is at this point, Cohen stresses, that modern science came into the world for good. According to him, so definitive and all encompassing was Newton's synthesis that "the fact of its triumph was by now as foreordained as anything in our fragile world can be" (Cohen 1994, p. 514).

Having completed his account of the birth of modern science, Cohen proceeds to evaluate the impact of the universe of precision on the medieval Aristotelian synthesis. According to him the tiny pocket of mathematized science inherited from the Greeks—Euclidean geometry, mathematical astronomy, Archimedean statics, and hydrostatics—underwent radical transformation and vast expansion after Newton. The area of natural philosophy that entered the universe of precision became greatly widened. As a result it led to the decline of Aristotelian natural philosophy—albeit over an extended period of time.[7] At the same time the religious worldview, which had been closely linked to Aristotelian notions and had given Europeans their sense of cosmological meaning, also found it necessary to yield to the universe of precision and create a space for itself outside it.

Cohen argues that intellectual factors alone, without the European Coloring, cannot have led to these changes. In his view, the intellectual factors that made the breakthrough possible in Europe around 1600 CE were also present in the Arabic–Muslim world. He writes:

> In structural terms, [with the Greek inheritance] we have here the same constellation of science as had obtained in the Muslim world: the adoption of the Greek legacy, its modest enrichment inside an overall framework left

intact, and the aggregate thus produced being complemented by some civilization-specific pursuits of its own making. (Cohen 1994, p. 509)

However, in the Islamic world no follow-up of any breakthrough would have occurred as in Europe because values perceived to be central to the faith were incompatible with modern science—or so argues Cohen. This leads him to propose, in the manner of Ronan, an inverse-Mertonian thesis that offers a diagnostics of the failure of Arabic science to develop into modern science. He is led to assume that something essential within Europe but absent within Arabic culture, namely the complex of factors he labeled the "European Coloring," made the crucial difference. This leads him to ask the following question:

> The "Islamic Coloring" hardly entered into productive interaction with the Greek corpus of science; the "European Coloring" did. Why was this so? Why was the Islamic Coloring so much poorer, *at least where science was concerned,* than its European counterpart? (Cohen 1994, p. 518). [Cohen's emphasis]

Cohen hardly recognizes the question itself as flawed. Indeed this question could be raised for all the cultures that inherited the Greek tradition—including the Hellenistic culture that created it and then stopped growing. One could ask, Why did the Hellenistic Greeks fail to develop modern science after an initial period of success, while Europe later did? Or why did the Byzantines, or for that matter medieval Europeans from the thirteenth to the sixteenth centuries, fail to develop modern science once they had inherited the Greek corpus? This could be followed by variant inverse-Mertonian theses blaming Hellenistic, Byzantine, or medieval European cultural and religious values.

Equally strange is the way productive interaction with the Greek corpus of knowledge is defined in Cohen's historical account. The Arabic civilization, which inherited the tradition, took it seriously, and developed it without inflicting violence on its fundamental structure, is charged with not having entered into productive interaction with it. The Europeans who initially rejected the corpus, and then made an effort to absorb it in the thirteenth and fourteenth centuries, only to proceed to subvert it in the sixteenth century, are considered to have engaged in productive interaction with it. If Cohen did not propose this notion of "productive interaction" so seriously, one might suspect that he is being ironical.

Moreover, if modern science developed as a result of not only Europe's productive interaction with the Greek corpus but also its dialogical interaction and assimilation of ideas and practices from multicultural sources—including the Arabic world—then Cohen's account of Arabic cultural values would become untenable. We would neither have the problem of explaining why Greek science never developed into modern science in any of the cultures that came to accept it, nor would we be inclined to consider Europe as engaging in productive interaction with it when it set out to subvert it. Let us therefore turn to some dialogical interpretations of the rise of modern science to see if they can lead us out of the problems into which Eurocentric approaches—such as Ronan's and Cohen's—seem to inevitably lock us.

Chapter 4

Multicultural Histories of Science

In recent years there have been numerous calls to reconsider the historical narrative of modern science as solely (or mainly) rooted in the Western or European tradition. This narrative has come to be suspect as historians examining traditions of natural knowledge outside Europe before modern times have increasingly come to recognize the contributions of non-European cultures to modern science. A major pioneer in this direction was Joseph Needham, with his series *Science and Civilization in China,* the first volume of which was published in 1954. Needham argued that many scientific ideas and technological discoveries earlier attributed to Europe had actually originated in China. Needham's groundbreaking studies—the most comprehensive modern survey of the scientific and technological accomplishments of any civilization outside Europe—were followed in 1968 by Nasr's *Science and Civilization in Islam,* which, although concerned only with documenting Arabic science on its own terms, nevertheless examined the profound influence of Arabic scientists on their modern counterparts. Shortly thereafter, Bose, Sen, and Subbarayappa attempted to do for Indian civilization what Needham and Nasr had done for Chinese and Arabic cultures. Their study in 1971, *A Concise History of Science in India,* constitutes a comprehensive survey of the main achievements of classical Indian science and the contributions it made to modern science. More recently, Martin Bernal, in his *Black Athena: The Afro-Asiatic Roots of Classical Civilization,* whose first volume appeared in 1987, argued that Greek civilization—generally considered by Eurocentric historians to be the sole ancient foundation for modern science—was profoundly influenced by the traditions of ancient Egypt and the Levant.[1]

These multicultural influences on the construction of modern science traced by Needham, Nasr, Bose et al., and Bernal have been made mainly from the point of view of the impact of one specific culture—Chinese, Arabic, Indian, or Egyptian—on European science, but taken together, they open the door to the possibility that modern science itself is a phenomenon with much wider multicultural roots than hitherto suspected. However, this multicultural turn itself may have been inspired by attempts made much earlier to trace the roots of modern science to medieval European culture. At the beginning of the twentieth century the French philosopher Duhem argued that modern science was crucially influenced by the scholastic tradition of the fourteenth century. Prior to Duhem it was customary to assume that the birth of modern science occurred with Copernicus, Galileo, Kepler, and others, who liberated thought from the servile subservience of scholasticism to Aristotelian doctrines. Going against this dominant view Duhem wrote:

> When we see the science of Galileo triumph over the hard-headed Peripateticism of a Cremonini [one of Galileo's adversaries in Padua], we believe, badly informed as we are about the history of human thought, that we are witnessing the victory of young modern Science over medieval Philosophy and its obstinate parrotry. But in reality, we are watching the triumph, prepared long in advance, of the science which was born in Paris in the 14th century over the doctrines of Aristotle and Averroes, which in the meantime had been restored to honor by the Italian Renaissance.[2]

By tracing the beginnings of science to the fourteenth century, Duhem put into question the key assumption behind the so-called "War of the Ancients and the Moderns" that had characterized previous historical writing—namely that while the moderns broke away from the ancients, the medieval scholastics slavishly imitated them. The end of this intellectual war was seen as the triumph of the modern science over the Hellenistic tradition. By tracing the roots of modern science to the scholastic heritage, Duhem not only proposed that modern science shared continuity with the scholastic tradition but, since the scholastic tradition saw itself as an attempt to come to terms with Greek thought, he made it possible for others to link modern science with the Greek corpus of knowledge. Many historians of science who followed him were to take up this theme and develop it much further. With the writings of Koyré, Dijksterhuis, and Burtt, we reach the current dominant view that Greek science was a precursor, rather than an obstacle, to modern science.

There was also another unintended consequence of Duhem's account—one he never addressed or even considered as a possibility. By locating the birth of modern science in the medieval era, Duhem led historians into a period in the history of Europe when it became connected to the more intellectually and technologically advanced Arabic and Chinese civilizations. At this time the Arabic–Muslim world stretched along the whole of the southern borders of Europe and also ruled European territories in the Iberian Peninsula in the west. The Chinese civilization also became linked to Europe through the territories of the Mongol Empire—an empire that included Russia in Eastern Europe. Sandwiched between the Arabic and Mongolian worlds, with territories in Western Europe ruled by Arab-Muslims and in Eastern Europe by Mongols, Europe was more open to external cultural influences than at any time since the Persian Empire ruled the Ionian Greeks during the lifetimes of Socrates, Plato, and Aristotle.[3]

By drawing attention to the role of medieval Europe in the genesis of modern science, Duhem prepared the way for Needham's and Nasr's histories, in which the impact of Chinese and Arabic sciences on Europe are first traced back to this seminal era. Moreover, Duhem made it possible to entertain the notion that if medieval science—with its religiously motivated philosophy, its organic metaphysics, its theory of five elements, its belief in qualities like hot, cold, wet, and dry—could serve as the precursor of modern science, then Chinese and Arabic cultures with a similar science of elements and qualities could also have played a role in the birth of modern science. He opened the door to the possibility that modern science might have a history that went further into the past than hitherto suspected—it did not simply emerge de novo in the sixteenth and seventeenth centuries. Thus the new historiography that traces the contributions of the Chinese, Arabic, Indian, and Egyptian traditions of knowledge to modern science can be considered as only taking further, into a wider geographical arena, Duhem's attempt to trace the roots of modern science to medieval Europe.

The most recent multicultural approaches to the history of modern science combine the earlier studies on the separate contributions of the Chinese, Arabic, Indian, and Egyptian traditions into more comprehensive accounts that trace the roots of modern science to the interaction of a plurality of non-European cultural influences. Although Needham did not himself undertake this task (focusing mainly on Chinese contributions), he pointed the way to such an approach when he wrote:

> It is necessary to see Europe from the outside—to see European history and European failure no less than European achievement, through the eyes of that large part of humanity, the peoples of Asia and indeed also of Africa. (Needham 1979, p. 11)

Needham is calling into question Eurocentric constructions of the history of science. There are those who would interpret this as tarnishing the repute of science by linking it to mythical, pseudoscientific traditions of knowledge from non-European cultures. Since science, as the Indian psychologist and culture critic Ashis Nandy has argued, is the dominant ideology today, this can also be easily interpreted as calling into question the unique status of modern science. Hence calls to transcend Eurocentrism tend to be perceived as extremist positions—and often evoke equally extreme counter-reactions. In his study *Eurocentricism*, Samir Amin writes:

> Resistance to the critique of Eurocentricism is always extreme, for we are here entering the realm of the taboo. The calling into question of the Eurocentric dimension of the dominant ideology is more difficult to accept even than a critical challenge to its economic dimension. (Amin 1989, p. 116)

In his study *Science and Technology in a Multicultural World*, the social anthropologist David Hess attempts to break this taboo by articulating a multicultural perspective on the history of science that traces the roots of modern science to non-Western cultures as well. Hess begins by questioning the conventional narrative of the Scientific Revolution of the seventeenth century told, as he describes it, "in the form of a dialogue between the Old Europe (ancient thinkers) and the New Europe (modern thinkers)" (Hess 1995, p. 63).[4]

He argues that this narrative ignores not only events prior to Old Europe but also those in the period intervening between Old and New Europe. According to him, during these eras, important exchanges took place between Europe and other cultures that are either completely ignored or only given a secondary plot status in Eurocentric narratives. Even though individual elements of the Eurocentric narrative, taken in isolation, do occasionally get questioned, such critiques are never brought together to subvert the overall Eurocentric structure of the story offered. They only appear as qualifying subplots designed to strengthen the Eurocentric main plot by acknowledging contributions that cannot avoid being recognized.

Hess refers to Needham's studies that show China to have been more advanced technologically than the West until the sixteenth century, and documents the numerous transfers of Chinese science and technology to the West in the medieval and later periods. Referring to one of Needham's lists—which includes magnetic science, equatorial celestial coordinates, quantitative cartography, the technology of cast iron, essential components of the reciprocating steam-engine, the mechanical clock, the boot-stirrup, the efficient equine harness, gunpowder—he argues that much of the technological infrastructure of modern science rests on borrowings from China (Hess 1995, p. 64). Mention could also be made of many other Chinese inventions such as weight-driven and water clocks, glass-making techniques, orgival (Gothic) architecture, water-raising machines, and paper making.[5]

Hess also draws our attention to the Arabic contribution. According to him, this influence was not only technological but also methodological and theoretical. He refers to the discovery of the lesser circulation of blood by al-Nafis as a possible influence on Harvey's discovery of the larger circulation. Moreover, following Butterfield, he argues that Harvey's experimental approach to biology was inspired by Averroes (Ibn Rushd, 1126–1198 CE),[6] who taught a secular and critical orientation toward the study of nature and medicine. Averroism (or at least Aristotelianism as seen through Averroes) was an important influence at the University of Padua when Harvey studied there (Hess 1995, p. 65).

Galileo was also a student at Padua and, moreover, did much of his work there. He too adopted the Averroist secular and critical approach to science. Moreover, Galileo was also influenced by the Arabic scientist Alhazen (Ibn al-Haytham),[7] known for his experimental approach to science, since he used the Alhazen theory of optics to show that the moon was not a polished mirror. Like Galileo, the mathematical astronomer Kepler also studied Alhazen's optics. Hence it is reasonable to suppose that both Galileo and Kepler could have been inspired by Alhazen's general approach to the practice of theoretical and experimental science (Hess 1995, p. 66).

Apart from optics, Hess also refers to Saliba's studies on the Maragha School of astronomy that flourished in the thirteenth and fourteenth centuries. Saliba argues that there are many parallels between the models of planetary motion and mathematical techniques deployed by Copernicus and the Damascene astronomer al-Shatir, which strongly suggests that Copernicus was aware of the latter's studies.[8]

Hess also considers the numerous and complex multicultural exchanges that took place in mathematics among different cultures over long historical periods. These led to significant mathematical achievements that were inherited by Europe and played a crucial role in the modern Scientific Revolution. There were the mathematical contributions of Egypt and Mesopotamia to Greek science, and the Arabic synthesis of the geometrical tradition of the Greeks with the algebraic and arithmetical traditions of Babylonia, India, and China. Hess also considers the influence of the Kerala School of mathematics in India during the medieval period that approached discoveries close to the calculus, although he acknowledges that the question of whether they influenced Europe continues to generate controversy. He concludes that cultures from Egypt and Mesopotamia to the Arabic world, India, and China made important contributions to the mathematical tools that made modern science possible.

After tracing the many multicultural influences on Europe, Hess proceeds to draw a number of highly dubious conclusions. First, he questions the notion that there was a Scientific Revolution in Europe at the dawn of the modern era. He thinks that the continuities between modern science and other traditions that preceded it make the notion of revolutionary break implausible. In this regard his views parallel those of historians Butterfield, Shapin, and Dear, although his conclusion is based on continuities within modern science and multicultural, not just medieval European, traditions.

Hess adopts the continuity thesis precisely because he thinks postulating a radical break separating modern science from earlier traditions would be ethnocentric, since it would marginalize the contributions of other cultures—especially the Arabic. However, a multicultural history of modern science acknowledging non-Western contributions can be written without denying that there was a revolutionary break involved in the emergence of modern science. Even if modern science was forged out of ideas, methods, and technologies developed in multicultural contexts, it is a sufficiently radical and unique achievement to be rightly described as revolutionary. Consequently, it is hardly ethnocentric to claim that a scientific revolution occurred in Europe—unless one denies the plausible claim that modern science first emerged in Europe. What would be ethnocentric is to deny the dialogical contributions that made the revolutionary break possible.

Equally questionable is the second conclusion Hess draws from his multicultural approach. He argues that we should replace the term "Western science" by the term "modern science" or "cosmopolitan science" since it absorbed contributions from many cultures. Yet somewhat inconsistently, he also wants to treat modern science as Western "ethnoscience" so that it cannot be used to ignore the valuable knowledge still carried by non-Western cultures, especially their environmental and medical knowledge (Hess 1995, pp. 67–68). His ambivalence appears to be related to the dual perspective he wants to adopt with respect to modern science: by virtue of its multicultural roots it is cosmopolitan; by virtue of its limitations relative to non-Western traditions, it is Western. Referring to modern science as "Western science" would be to ignore the contributions of other cultures that made it possible; referring to modern science as "cosmopolitan science" would marginalize other traditions of knowledge by implying that they have been completely displaced by this science synthesized in the West.

The aforementioned difficulties that Hess confronts can be easily resolved if we make a distinction between premodern and modern traditions of science. Then we can treat modern science as Western science—should we want to—provided we recognize its multicultural roots. Since we are prepared to refer to the science that developed in the Arabic-speaking world as "Arabic science" in spite of its roots in Greek, Persian, Indian, and (to an extent) Chinese traditions, there is no reason to deny the term "Western science" to the tradition that developed in the West in spite of its roots in other traditions. The universal or cosmopolitan status of modern (or Western) science can also be left as an open question. It is quite possible that, in effecting the synthesis of modern science from premodern traditions, Europeans created a science that has obscured our understanding of some areas of reality now understood better by other traditions of science. It may well be correct for many environmentalists, feminists, and multiculturalists to turn to what they think are the more holistic or organic perspectives of premodern traditions to remedy the blind spots they perceive in modern science. If they are right, then modern science cannot legitimately claim to be universal science.

Indeed the historical question of the roots of modern science and the epistemological question regarding its universal scope or cosmopolitan status are intimately connected. If modern science has roots in Europe alone, and grew independently of other traditions, it lends credibility to its universalist and cosmopolitan claims—it is quite possible that a universal science emerged once in Europe through the discovery of how to discover, that is, through the finding of the one and only scientific method, and then spread elsewhere. However, if modern science drew on earlier traditions of science, then it is conceivable that a new order of science may yet emerge in the future by drawing on modern science and other premodern traditions. This would lead us to be wary of claims that modern science is both universal and cosmopolitan and that we can discount the claims of other traditions to reveal aspects of nature ignored by modern science. Thus, whether we take a Eurocentric or a dialogical approach to the birth of modern science would have significant implications for our evaluation of the future relationship between traditional cultural reservoirs of knowledge and modern science.

Chapter 5

Toward a Thematic Approach to Multicultural History

The multicultural histories of science have also been heavily criticized. Although most of these have been directed against Needham's account of the impact of Chinese science on modern science, they could easily be extended to studies of other cultural influences. One major objection against such dialogical histories is that they fail to make the important distinction between science and technology. This allows them to pass off technological contributions as contributions to science, so that it appears as if a case has been made for multicultural influences on science when, in fact, a case has really been made only for multicultural contributions to technology. For example, the medieval science historian Lynne White has argued that if one removes all premodern transmissions of various Chinese technical inventions to Europe, then nothing remains to substantiate Needham's claim that Chinese science made significant contributions to modern science. White writes:

> Needham has described a considerable group of technological innovations that either clearly or possibly reached Europe from China before the Portuguese first sailed around Africa to Asia. I cannot recall, however, any significant scientific theoretical concept coming from China to the West even when such concepts in China can be shown to antedate identical or similar ideas found in Europe. (White 1984, pp. 178–179)

White's assumption that Needham has only made a convincing case for Chinese technological contributions is questionable. Needham also has presented arguments to suggest that Chinese views on cosmology, observational astronomy, pharmacology, and the magnetic sciences—even philosophy, via Leibniz—played a crucial role in shaping the evolution of ideas in modern science. While these claims are controversial, they are not claims that can be reduced to mere technological influences.[1]

White could concede this point but nevertheless argue that Needham claims multicultural influence simply on the basis of the existence of parallel ideas in China before they emerged in Europe, without giving any additional support for his claim that this is the result of cross-cultural transmissions. The mere fact that an idea emerged earlier in China does not mean it could not have been discovered independently in Europe later. Indeed Needham himself concedes that his claims for transmission are often based simply on the observed fact that an idea or practice occurred in China before it appeared in Europe. He vindicates this approach by arguing, without convincing his critics, that "the details of any transmission are difficult to observe" (Needham 1969, p. 83).

This controversy raises a number of questions. In the absence of any direct evidence for transmission, can we legitimately infer that a discovery first made in one culture influenced another culture in which it later emerged simply on the basis of the fact that a corridor of communication existed between the two cultures that would have rendered this transmission possible? Or, should we require the more stringent condition that a claim of influence can be considered plausible only if, over and above the possibility of influence, we also have evidence for actual transmission of ideas? Given Needham's observation that the details of a transmission, even if it had occurred, are difficult to observe, such a requirement may be too stringent. However, inferring influence merely on the basis of its possibility may be a requirement that is too weak.

Nevertheless, we have to make a decision, one way or another, in order to write history. We seem to be forced to choose between one of two ways of establishing transmission—one that is too stringent and another that is too weak. Moreover, the decision we make, when systematically adopted, will profoundly shape the history we write. If we take the weak requirement that priority of discovery and the existence of a corridor of communication are crucial, we will assume transmission to have occurred whenever it is possible. We are likely to be led to a dialogical history of modern science simply because many of its discoveries have been anticipated by non-Western cultures and came to be made in Europe only after contact with these cultures. By contrast, if we take the strong view that absence of evidence for transmission indicates independent discovery, we are likely to arrive at a Eurocentric history since, as observed by Needham, details of transmission are hard to find.

Needham himself chooses the dialogical alternative because he deems the burden of proof to lie with those who stake a claim for independent discovery. He writes:

> Of course there may have been some degree of independence in the European advances. Even when we have good reason to believe in a transmission from China to the West we know very little of the means by which it took place. But as in all other fields of science and technology the onus of proof lies upon those who wish to maintain fully independent invention, and the longer the period elapsing between the successive appearances of a discovery or invention in two or more cultures concerned, the heavier that onus generally is. (Needham 1970, p. 70)

Thus Needham assumes that it is natural to presume cross-cultural transmission under conditions where this is possible, unless we can demonstrate independent discovery. The historian of science Cohen disputes this principle adopted by Needham. He argues that although Needham rightly recognizes that the absence of evidence for transmission does not prove independent discovery, he wrongly concludes that absence of evidence against transmission allows us to assume influence simply on the basis of its possibility. Cohen writes:

> Although it is true that, if no transmission took place, there will necessarily be no source material to indicate transmission, the absence of source material need not necessarily prove independent rediscovery, even though it would seem to count against transmission rather than for it. Thus the two possibilities are not logically symmetrical. But since one can argue endlessly over whether the affirmation that stands in need of proof is transmission or independent rediscovery without ever getting anywhere, it would seem better to seek another criterion if one wishes to say anything at all about these really quite doubtful yet highly important matters. (Cohen 1994, p. 436)

However, when it comes to his turn to write history Cohen does not offer us any new criterion. Instead he himself adopts the strong criterion of direct evidence for transmission by arguing that absence of evidence for transmission should be taken to count more against transmission than for it. This ultimately makes him slide into a Eurocentric history of science— one that assumes independent discovery in Europe of many ideas anticipated earlier in China. If he had followed Needham and assumed cross-cultural influence as the default explanation for the existence of parallel ideas, his historical narrative would have been quite different.

The radical influence of an initial choice we make on the final picture we see parallels our experience of ambivalent gestalt images. Consider the duck/rabbit figure psychologists are familiar with. Our experience of the figure is crucially influenced by the single decision we make: whether what we see is a bird or a mammal. Similarly our historical narratives also have very different outcomes depending on how we wish to interpret the existence of parallel ideas in two cultures in contact—as the products of dialogue or independent discovery. Does this mean that ultimately the choice of a dialogical or Eurocentric history has to be arbitrary? Should we therefore adopt an agnostic standpoint that chooses neither the Eurocentric path taken by Cohen nor the dialogical one adopted by Needham? This would leave us in a permanent historical limbo; worse, it may favor Eurocentric histories simply because they can be written by taking the presence of a discovery in Europe as a fact, but treating as disputable its non-European origins. Hence there is no agnostic standpoint.

Cohen also raises a second objection to Needham's historical approach. He argues that it exaggerates the Chinese contributions to modern science even if it were the case that Chinese ideas and practical techniques were actually transmitted to Europe. Hence he rejects Needham's estimate of the value of Chinese contributions to modern science:

> [They are] heavily flawed on several counts, of which the most important are the absence of sources that even begin to point at transmission and the consistent aggrandizement of what Chinese finds, even if passed on indeed to Western Europe, may have meant for early modern science in the best of cases. (Cohen 1994, p. 437)

The charge of aggrandizement of claims is significant. Moreover, Cohen imputes Needham with exaggerating the impact of the Chinese achievements in order to compensate for European claims that he considers tilted in the opposite direction. There is no doubt that Needham considered it an important part of his moral mission to put down European arrogance. He considered such arrogance to be the result of inflated claims to originality that had ignored Europe's indebtedness to discoveries from other parts of the world in creating modern science (Cohen 1994, pp. 437–438). However, while Cohen concedes an admiration for Needham's moral objectives, he argues that we cannot let morality decide matters of historical fact. Since it is possible to trace many of the component elements that came together in Europe to create modern science into historical and intellectual developments within Europe itself, there is no basis for attaching significance to external cultural influences. By doing so Needham aggrandizes the claims of non-European cultures, in particular China, by inflating its influence on the development of ideas and notions already explicable as internal developments of ideas within Europe.

However, by charging that many of the ideas that Needham imputed to Chinese influence can be traced to European sources, Cohen overlooks an important point. It is the case that all cultures—especially the civilizations we are here concerned with, such as the European, the Arabic, the Indian, and the Chinese—have sufficient resources within themselves to treat many ideas that developed elsewhere, and influenced them, as articulations and elaborations of notions and techniques within their own tradition (although these notions and techniques had existed only as marginal or recessive themes in them). This means that even if they do undergo intellectual mutations under the impact of other cultures, it is easy for them to deny the significance of such influences by rooting the changes into their own traditions.

For example, the atomic idea emerged in China with the Mohists. Mo Tzu (c. 471–389 BCE) was an older contemporary of Democritus (c. 460–370 BCE). Assume for the purpose of argument that Mo Tzu, being over a decade older than Democritus, had proposed the atomic theory earlier. Then, clearly, he had priority of discovery of the theory. How then are the Chinese today to write the history of the atomic theory they teach in their schools? They can adopt one of two routes—they could say that the modern atomic theory is an elaboration of the Mohist atomic theory and write history from this Sinocentric point of view. This seems unfair to the European contribution, since the only reason that the atomic theory is taken seriously in China today—where it existed only as a minor theme in traditional Chinese thought—is the influence of scientific ideas from the West. To write the history of contemporary atomic ideas in China as emerging from Mohist ideas would be a form of aggrandizement.

Yet this sort of historical reconstructions are often made when cultures confront alien influences—especially cultures sufficiently rich in possessing a wide variety of conceptual resources. Dominant and elaborately articulated ideational themes in an alien tradition are absorbed but treated as internal articulations of previously undeveloped minor themes within the recipient culture. The classic example in Chinese civilization is the development of neo-Confucian thought, which articulated a new metaphysics largely under the impact of dominant ideas in Buddhist thought but, since Confucians perceived Buddhism as an alien tradition, interpreted the change as a natural development of what had been only minor notions in the earlier Confucian heritage. Nevertheless, without the impact and example of Buddhist metaphysical ideas, these recessive Confucian themes would not have been amplified into the full-fledged metaphysics they generated—nor could they have been. The reason we assume that neo-Confucians were influenced by Buddhist thought, even though they are able to trace the roots of their ideas to their own tradition, is that some dominant themes in

Buddhist thought also became dominant in neo-Confucian philosophy after contact with Buddhism, although these same notions existed as minor themes in earlier Confucian thought. We do this because the change in Confucian notions, when we shift from classical Confucianism to neo-Confucianism, could not even have been entertained seriously but for the exemplar provided by Buddhism. The same reason that leads us to think that the atomic theory taught in Chinese schools is an elaboration of modern atomic theory, and not the Mohist theory, must make us conclude that Chinese neo-Confucian ideas were elaborated under the impact of Buddhism—even though they may also have transformed these Buddhist ideas in the process. To deny the influence of Buddhism on neo-Confucianism, in this regard, is a form of aggrandizement of the role played by minor themes in the Confucian tradition while repudiating the role played by major ideational themes in Buddhist thought.

Indeed it may be argued that well-developed civilizations, such as the European, Arabic, Chinese, and Indian, have sufficient resources within themselves to absorb any ideational themes they find attractive in another culture as an elaboration of a minor theme already present within their own culture. Thus they would always be able to construct an immanentist history of any influence from the outside. These major cultures have enough cultural resources to insulate themselves into separate historical enclaves even as they dialogically influence each other, by erasing historical acknowledgment of such influences—and treating their intellectual histories as autonomous internal developments of what were earlier recessive themes within their own respective cultures.

Such historical constructions have unfortunately become much too common. In many postmodern histories we see implausible attempts to argue that the main features of modern science, including its latest theoretical discoveries, were really invented in the Arabic world, anticipated by ancient Indian seers, discovered by Chinese sages, and so on. Such constructions of history are made possible by selectively tracing the dominant ideas of modern science to marginal, recessive notions within the selected premodern tradition. This is generally done without regard to the following question: Would the recessive notions in the particular tradition have become identified as significant, and combined in the ways they now are, without the guidance and template provided by the dominant ideas of modern science developed in Europe?

Yet the blame for such one-sided tendentious histories that aggrandize the role of minor cultural ideas from non-Western cultures cannot be isolated from the same tendency on the part of European historians. Indeed they may be deemed counter responses to the injustice perceived in standard Eurocentric histories of science. European historians themselves have generally ignored the contributions that major concepts and principles from other cultures made to modern science by masking their influence and tracing their presence in current science to minor themes in Hellenic or medieval European thought. Such histories of modern science are equally immanentist. The only difference is that Eurocentric immanentism is better placed to project itself onto the global stage by virtue of the political, economic, and military dominance of the West.

Indeed, during the "Axial Age," a term coined by Jaspers to refer to the era that led to the birth of civilizations in the Mediterranean, India, and China from about the sixth century BCE, there was an explosion of intellectual ideas in all these areas that created a storehouse of concepts that anticipated, in one way or another, many notions in modern science. These ideas provide sufficient material for reconstructing the history of modern science as simply articulating ideational themes already anticipated in each of these civilizations. Traditionally it has been customary to treat the Hellenic civilization as providing the intellectual matrix from which history has to be reconstructed—largely because it is closely identified with Western Europe, where modern science emerged. However, today such accounts are no longer taken for granted—ancient Egypt and Mesopotamia, Arabic civilization, as well as Chinese and Indian cultures have all been considered to have shaped the rise of modern science. Since many of these cultures have interacted with Europe in the past, and each offers enough variety of conceptual resources to root the ideas of modern science as emanating from the influence of ideas within itself, it lends scope for endless debate on the roots of modern science, and opens the door to the possibility that there may be no resolution to the historical question being addressed: Is modern science a multicultural or European phenomenon? Or a dialogical product that can be seen as both by virtue of who contributed to it and where it was born?

Clearly we cannot decide if an influence occurred by simply asking whether certain ideas in modern science can be found in the premodern ideas of any particular culture. What is crucial is to decide when an idea from outside Europe can be said to have influenced European thinkers even when the same idea may also have been proposed earlier by a thinker within Europe. If the idea had not been taken seriously in Europe until its significance was shown by contact with the culture outside, we have reason for presuming influence. If an idea remained a recessive theme in Europe before the emergence of modern science, and its significance was not perceived until Europeans confronted it as a dominant theme in another culture, which then led them to articulate it into a major theme in modern science, then we have to concede influence. It is reasonable to assume that the rise of the conceptual theme in modern science was the result of the influence of the non-European culture. Adopting this strategy allows us to circumvent the limitations of not only Needham's weak criterion of transmission that assumes priority and possibility of transmission to imply transmission, but also Cohen's strong criterion that establishing transmission requires us to present direct, independent evidence—a condition so stringent that it biases history toward a Eurocentric direction.

To pave the way for such a new criterion I would like to look more closely at the notion of ideational themes. The notion itself is derived from the historian of science Gerald Holton, who used it to account for both stability and change in the evolution of scientific thought. In his book *Thematic Origins of Scientific Thought*, Holton argues that scientific thinking does not only appeal to empirical and analytic (mathematical) propositions but is also guided by a third class of beliefs he refers to as thematic principles. Such thematic principles, or themes, are in many cases drawn from natural philosophy and may include methodological, ontological, and conceptual principles. Themes often persist in the background (even in scientific discourse) by guiding and inspiring the articulation of scientific theories. However, being neither verifiable nor falsifiable, they cannot be deemed to be scientific beliefs in themselves. Nevertheless, they constitute the nucleus from which empirically testable theories can be articulated. According to Holton, many of the major revolutions in scientific thought can be illumined by a thematic analysis in which particular thematic principles originating in natural philosophy can be seen to develop over time into full-fledged scientific theories with an associated empirical basis.

The sociologist of science, Susantha Goonatilake, has recommended that we deploy Holton's idea of themes to show how multicultural reservoirs of knowledge can be exploited to advance future science. He writes:

> Holton (1973) has pointed out that from Greek times the theories of science have been built up by a small collection of what he calls themes and anti-themes. Examples of these are complexity-simplicity, reductionism-holism and continuity-discontinuity. If these constitute some of the ultimate building blocks of existing theories of science, further themes and anti-themes or derivatives thereof could be found from other cultures. This would give a larger thematic mix which would provide for a greater variety of building blocks for new concepts and theoretical frameworks.[2]

Goonatilake's recommendation may also be applied to the history of modern science by seeing how themes and antithemes from non-Western cultures influenced science. We have seen that many developed cultures have the same broad complex of themes and anti-themes (which really constitute another kind of theme). However, they may not have developed every one of them to the same extent. A particular thematic idea may have undergone extended development in one culture but not in another. For example, atomic theories were developed extensively in Indian but not in

Greek culture; deductive reasoning was systematized in Greek culture, but correlative thinking in Chinese. Hence it is possible to suppose that Europeans in the modern era, by coming into contact with many cultures, were able to fuse together thematic principles that had been greatly developed in other cultures with developed themes already present within their own culture to create modern science. Nevertheless, since the developed themes taken from other cultures already existed as minor themes within Europe, they could easily write a history of the growth of modern science that traced the combination of thematic ideas they created for modern science as solely rooted in the European tradition. In effect they could refuse to acknowledge the impact of external traditions in spite of the fact that without this external impact they would never have recognized the potential of many developed thematic principles they now trace to recessive themes in their own tradition.

We will now propose a criterion that does not require the strong proof of transmission required by Cohen and the weak proof of priority allowed by Needham to decide whether a case for cross-cultural transmission of an idea exists. The transmission criterion we propose is as follows:

> If, shortly after a new corridor of communication opens between a culture A and a culture B, and great interest shown by A to understand B, a theme becomes dominant in A similar to a dominant theme in B, then we can presume that the development of the theme in A was due to the influence of B, even if the new theme had existed as a recessive theme in A prior to contact between the cultures.

It is important to emphasize that the above thematic criterion is not an attempt to identify unambiguously every case of cultural transmission. Indeed there could be cases of transmission that do not meet the criterion. For example, a culture may be influenced by another culture even when the corridor of communication between them has not changed dramatically for a long time—it is not a *new* corridor. It may be a response to new conditions in A that lead it to become more receptive to a thematic notion in B. It is also possible for a culture to influence another without any motivation on the part of the receiving culture to be so influenced (as the influence of colonial powers on colonized societies testifies). Finally, a culture may be influenced by a minor theme in another culture, which the latter did not consider to be an important element in its tradition. All these cases of influence do not meet the thematic transmission criterion proposed. Thus the criterion does not furnish necessary conditions for presuming influence; that is, it does not identify every case of influence.[3]

What it does provide is a sufficient condition for presuming influence, that is, any case that meets the condition can be deemed a likely case of cross-cultural influence. For, if a culture A finds a new corridor of communication between itself and another culture B, becomes greatly interested in the ideas and practices of B, and a short time later develops thematic principles within itself that reflect the dominant thematic ideas of B (albeit often in a transformed or modified form), then it is reasonable to assume that these thematic ideas emerged as a result of the contact of A with B. Moreover, even if it were known, or discovered later, that the same thematic idea exists in a recessive and underdeveloped form within the culture A, this cannot be deemed evidence that A was not influenced by B. It is only evidence that A had anticipated the theme without being either willing or able to develop it further until the influence of B.

Let us now compare the proposed criterion with the weak and the strong criteria adopted by Needham and Cohen, respectively. Unlike Needham's weak criterion, it does not require us to assume that priority of discovery and possibility of transmission are sufficient to imply influence—although they are necessary conditions for claiming transmission. To claim transmission we must also show that the new theme developed shortly after a corridor of communication opened between two cultures, and that the new idea was a dominant theme in the influencing culture, but not in the culture that was influenced. These additional conditions to the ones already proposed by Needham render the criterion stronger than his criterion for claiming transmission. However, the criterion does not require evidence in the form of documents, explicit acknowledgments, or other direct proofs. In this regard it is less stringent than Cohen's criterion, which is loaded in favor of Eurocentric historians.

In this study we find that Eurocentric histories of science are the result of systematic constructions that violate the thematic criterion. We will see that a great deal of historical evidence supports the view that Europe has repeatedly shown strong interest in understanding ideas in other cultures and their scientific traditions shortly after a corridor of communication opened Europe to them in modern times. Moreover, dominant themes in these cultures became dominant in European thought only after this encounter, although they may have existed as recessive thematic notions in Europe. However, once they became internalized as dominant themes in European consciousness, historical genealogies were constructed to reinterpret them as autonomous developments of the minor themes that most closely resembled them within the European tradition. In this fashion, a history of intellectual changes in Europe came to be constructed that ignored the multicultural influences on Europe and the dialogical processes through which Europe appropriated these ideas from the outside. Such a historical construction could only be sustained because the rich heritage of Axial Age Greek thought (like that of Axial Age India and

China) provided a wide and diverse-enough spectrum of thematic ideas to make this possible. Nevertheless, the historical account itself has to be deemed one-sided and distorted because, without the external impacts on Europe, these internal intellectual changes are hardly likely to have occurred.

Nothing in this account should be taken to imply that the synthesis achieved by Europe was not a unique cultural achievement. It was distinctive precisely because it drew upon dominant themes from a wide number of cultures, over an extended period of time, to forge out of them a conception of nature and society that had never existed before in any of the cultures that provided the resources for this construction. The accomplishment is creditable to the genius of Europe, but was also based upon prior achievements of outside cultures from which Europe drew ideas and practices. The novel synthesis of modern science in Europe was founded upon multicultural traditions. By virtue of its place of birth, modern science can be described as Western; by virtue of the component elements that went into making modern science possible, it has to be deemed multicultural. Both descriptions are accurate because modern science emerged in Europe through its dialogue with ideas found in other cultures.

Chapter 6

What Made the Renaissance in Europe?

The European Renaissance is often considered to be the crucial era that led to the emergence of the modern world. The standard view of the Renaissance is that it was the rebirth in Western Europe of Hellenic and Hellenistic philosophy, literature, arts, and sciences after a long intervening period following the collapse of Rome and the rise of scholastic philosophy in the thirteenth century. During the intervening period—often characterized in the past as the "Dark Ages," although more recent histories tend to see important developments in this period that paved the way for the future course of European history—this heritage is considered to have been carried by the Arabic civilization, which, although adding to it in minor ways, did not develop it too far. When this fossilized heritage was returned to its rightful cultural progeny in Europe, it became revitalized once more and led to the intellectual revolution that produced modern science and philosophy.

However, this account minimizes the added corpus of knowledge Europe received from Arabic civilization when it entered the Renaissance. In the first place, although this view recognizes that Greek science was reborn in Europe, it fails to see that the totality of the sciences transmitted far exceeded anything known to the Greeks. These include a far more advanced number theory and algebra, a new system of trigonometry, a medical corpus much greater than that available in the Greek world, and an entirely original theory of optics more powerful than anything known to the Greeks and that was not only to form the mathematical basis for Renaissance art but also to inspire new directions in scientific practice. This excess renders highly questionable the standard view that Arabic culture had merely served as the repository of Europe's Greek heritage.

Second, the standard account also overlooks the fact that European scholastic philosophy benefited from the intense debates among Arabic thinkers who were concerned with reconciling science and revelation. These debates had led to a profound division in Arabic philosophy between those who followed Greek rationalism, the *falsafah*, and others more inclined to prioritize theological concerns, the *kalam*. Both the *falsafah* and the *kalam* engaged in controversies among themselves, and with each other, on ontological questions about whether the world was eternal or created in time; epistemological issues about the existence of necessary causes in nature; and theological concerns regarding the relationship of reason with revelation. Since Catholic thinkers shared many religious ideas with Muslims that were not part of the Greek heritage—belief in a monotheistic God; in the creation of the world ex nihilo; in resurrection, prophecy, providence, the Last Day; and so on—the approaches adopted by Arabic thinkers in dealing with these issues were to profoundly affect the scholastic tradition in Europe.[1] Arabic philosophical works translated into Latin played a seminal role in the growth of medieval European thought—especially the studies of Maimonides (1135–1204 CE) and Averroes.[2]

Arabic art also influenced Renaissance art. This might appear surprising in the light of the fact that the aesthetic tradition of Arabic civilization was largely nonrepresentational—given the Islamic proscription of representations in general—while European Renaissance art is quintessentially representational. However, the influence here is not on the objects artists studied but the methods they deployed. Renaissance artists applied mathematics to art through theories of perspectivism; the application of mathematics was also a characteristic of Arabic art that, inspired by the mathematical mysticism of both *falsafah* and *kalam* thinkers, viewed abstract mathematical art as symbolizing the unity of God. Moreover, the perspective theories of Renaissance artists were profoundly shaped by the optical theory of the mathematical physicist and astronomer Alhazen, considered by most historians to be one of the greatest natural scientists of classical Arabic civilization. His theory had been translated into Latin in the twelfth or thirteenth century and became the dominant paradigm of optics in Europe—used by both Galileo and Kepler—until it was displaced by Newton's optical theory (which, as we shall see later, also grew in part out of this tradition).

Another external influence on Europe that gets overlooked in the standard view of the Renaissance as merely resurrecting the Greco-Hellenic heritage is the influence of China on the profound technological revolution in Europe during the same period. Prior to Needham's seminal studies on the transmission of Chinese technological discoveries to Europe through the corridor of communication created by the Mongolian Empire, this revolution had been assumed to have been an indigenous achievement whose sudden eruption required explanation. One explanation was that the ravages of the plague had so depopulated Europe, and created such a shortage of labor, that it motivated technological innovation to increase productivity—a case of necessity becoming the mother of invention. This explanation is partly correct because the plague, which entered Europe through the communication networks created by the Mongols, did inadvertently create a demand for new productive technologies. However, even though Europe experienced the need for more efficient instruments of production, this did not mean that it met the need by creating them de novo—instead it drew on a reservoir of Chinese inventions made over a period of fifteen centuries.[3]

The technological contributions of China to Renaissance Europe, first systematically documented by Needham, have not been denied—even by historians such as White and Cohen, who otherwise question the claims of Chinese scientific contributions made by Needham. According to Needham these technologies had multiplied and developed naturally in China in consonance with the Chinese organic materialist worldview over a long period of time extending over a millennium. By contrast they emerged suddenly in Europe. In the past, prior to Needham's seminal discoveries, the emergence of these new technologies in Europe had been attributed to vague notions such as European, or Germanic, genius for technology.

Moreover, it is highly likely that the mechanical vision which inspired many of the founders of modern science was itself conditioned by the impressive impact of these Chinese technologies, which entered Europe through the corridor of communication created by the Mongolian Empire that ruled from Russia to China in the thirteenth and fourteenth centuries. These discoveries occurred in China, as Needham maintains, because of the pragmatic orientation of the overall Chinese organic materialist worldview. Nevertheless, they did not lead to a full-fledged mechanical conception of nature as happened in Europe. It is possible that this did not happen in China because the technological discoveries arose in Chinese culture slowly over an extended period of time, so that each invention came to be absorbed and integrated into its wider organic perspective before making way for others. By contrast these mechanical discoveries arrived in Europe over a very short period of time and impressed the European consciousness sufficiently to trigger a transformation in thinking that reconstructed the universe in the image of a machine. The new vision in turn laid the intellectual, conceptual, and pragmatic basis for Europe to overtake China by developing and articulating even more sophisticated mechanical technologies.[4]

Hence, viewing the Renaissance as essentially the rebirth of Greek knowledge ignores the profound role of two other civilizations in shaping developments within Europe during this period. First, this view ignores the impact of the Arabic civilization that spanned Europe all along its southern borders and included parts of Western Europe in the Iberian Peninsula. Second, it overlooks the Chinese impact whose technological discoveries were passed onto Europe by a Mongolian empire that spanned its eastern front by ruling over parts of Eastern Europe in Russia. In a profound sense Leonardo da Vinci, one of the leading cultural figures of the European Renaissance, is the classic embodiment of these twin external cultural influences—the archetypal Renaissance man whose interest in the application of mathematics to art and the application of mechanical techniques in engineering, reflects the dual Arabic and Chinese influences on the Europe of his time.

Moreover, these influences were not the result of the passive diffusion of cultural ideas into Europe. Both the Mongol and Arabic rulers sought to draw Europeans into their arena of concerns; and conversely there were attempts by Europe to enlist these cultures into its political and social affairs. First, the Mongols went out of their way to recruit administrators and merchants from Europe, as well as the Arabic world and other parts of Asia, in order to dilute the Mandarin Chinese influence within China and to promote trade across the wide regions that had come under their control. Marco Polo was but an example of many other Europeans—albeit one who became famous because he recorded his travels—who had entered the Mongolian Empire to trade and perform administrative service for its rulers.[5] The Mongols also went out of their way to encourage dialogue and cultural contacts between people in areas under their rule and even those outside—for example, they organized religious conferences in their capital Karakorum attended by Islamic, Christian, Buddhist, and other scholars.[6]

The Europeans themselves saw the Mongols as allies against the threat of Islam. Their search for Prester John beyond the borders of the Islamic world, and the emissaries sent by the Pope to the Great Khan during this period, were designed to forge a partnership with the Mongols to deal with the real threat to Europe posed by the Arab-Muslims.[7] This was the time of the long-drawn-out war of the *Reconquista* in the Iberian Peninsula that ended only in 1492, when Granada was captured by Christian forces. The Crusades, which began in the eleventh century, were still in progress in 1453 when Constantinople fell to the Turks. However, these wars themselves did not prevent cultural interaction between the Arabic and European worlds. Just as the Mongols were prepared to recruit European and Arabic talent to run their empire, so too did Europeans recruit Arab-Muslim administrators in Norman Sicily and the Turks deploy Christian ones in the Balkans.

Such contacts opened Europe to the technologically and economically more advanced civilizations outside. The slow and gradual reconquest of territory in Western Europe in the Iberian Peninsula from Arab-Muslim rule widened European horizons by making increasingly available the works of Arabic scholarship, culture, science, and philosophy found in the captured libraries. The process of cultural assimilation was also facilitated by scholars well versed in the Arabic traditions who remained in Spain after conversion to Catholicism, and others from different parts of Europe who traveled to the newly reconquered territories to acquire knowledge. New universities in places such as Bologna, Padua, Paris, and Oxford, modeled on Arabic centers of learning, were set up. Openness to Arabic learning was supported at the highest levels. There was precedent for this in the great teaching master Gerbert of Aurillac (943–1003), who traveled extensively in Spain to acquire Arabic learning directly and attracted many students who spread his views throughout Europe. Gerbert was later to become Pope Sylvester II. The University of Padua, subsequently to become the alma mater of Copernicus, Galileo, and Harvey, adopted the curricula of Arabic institutions in its various courses (with the exception of theological and religious studies). These new institutions of learning became centers for disseminating Arabic philosophical, scientific, literary, and scholastic ideas throughout a receptive, albeit threatened, Europe.

The universities themselves were fed by a massive and sustained effort to translate Arabic works beginning with the end of the tenth century. Gerard of Cremona translated more than seventy works from Arabic between his arrival in Toledo in 1160 and his death twenty-seven years later. They included Ibn Sina's (Avicenna's) *Canon* and al-Razi's (Rhazes') *Liber Asmansoris,* both of which became canonical texts in European medical circles. He also translated two fundamental works on optics by al-Kindi, and Alhazen's *Optics.* The latter work, by treating vision as resulting from rays entering the eye through a "visual cone," with its apex at the eye, laid the basis for the laws of perspective in Renaissance art.

However, such translations were not merely the efforts of scholars who went to Spain. They were made from the entire span of Arabic civilization circling the Mediterranean. In Syria, during the thirteenth century, Philip of Tripoli translated the *Secret of Secrets* by al-Razi—a unique alchemical work, so untainted by the generally mystical notions that informed the discipline at the time that it has come to be seen as a protochemistry. From North Africa there were many translations by Constantine in the early eleventh century. It was also here that Leonardo of Pisa learnt about Indian numerals, knowledge of which entered Europe through the publication of his introduction to Arabic algebra in 1202.[8]

However, it was from the territories reconquered from the Arab-Muslims that the most important influences entered Europe. After Spain, the principal center for radiating Arabic influence was Sicily. This island had been ruled by the Arab-Muslims in the tenth and eleventh centuries before it was conquered by the Normans. Under the enlightened Norman rulers Roger II and Frederick II, Sicily continued to remain open to the Arabic world of learning and the cultivation of Arabic geography, astronomy, zoology, and optics were encouraged in the Palermo court (Goldstein 1980, p. 111). It was here that Michael Scot translated al-Bitruji's (Alpetragius') *On the Sphere*—a work that, according to Goldstein, played an important role later in liberating European astronomy from the over-shadowing influence of Ptolemy.[9]

Scot also translated the works of Averroes, who was to become an omni-present figure in medieval European scholastic debates. Averroism became a rallying point to liberate philosophy from excessive subjection to theology and was a significant movement at the University of Paris during the time of Thomas Aquinas (1225–1274 CE). The Averroists called for the separa-tion of religious and scientific truth—a separation we take for granted now but that was then perceived as subversive and heretical. Even Aquinas, who had been sent to deal with the Averroists, came to be charged with Averroist inclinations (although he was later reinstated after the charges were revoked). The influence of Averroism was to continue to be felt in the institutions of higher learning in Europe for over three centuries—it per-sisted as a continuous undercurrent that played an important role in the University of Padua during the times of Copernicus, Harvey, and Galileo.[10]

The work of the translators in bringing Arabic science and philosophy to Europe cannot be underestimated. Many of these translators interpreted their efforts as restoring to Europe the classical heritage of the Greeks—a view that came to dominate the ideological position adopted by Renaissance European thinkers. This interpretation was partly true because Arabic scholars themselves considered their scientific and philosophical tradition as deeply indebted to the Greeks—but not only to the Greeks since they also acknowledged the contributions of the Indians, Persians, Syrians, Egyptians, and others. However, the translators were working under difficult conditions, in the middle of a religious conflict between Islam and Christianity, and they were in no position to credit Arab-Muslim civilization too obviously. But this did not mean that they were not concerned with bringing into Europe the heritage of Arabic thought in science and philosophy precisely to defend it against the threat posed by Islam. In this respect they behaved no differently from the Japanese of the Meiji era who undertook a massive translation of European scientific and technological works to defend themselves against the "guns and ships" that had recently humiliated the Ching Empire in China.

By inventing the doctrine of the Dark Ages, in which Europe lost its Hellenic heritage now ostensibly being recovered in the Renaissance, the translators created an ideological shield behind which they could proceed unhampered. Such a strategy to restructure a threatened tradition, by absorbing influences from the tradition posing the threat, and interpreting it as a return to primal roots, is by no means unique to Renaissance Europeans. We find similar patterns of strategic appropriation and denial in other traditions. For example, the Hindu Vedanta doctrines were structured under the influence and threat of Buddhist domination in India but the process was treated as a return to the original Vedic sacred scriptures. The neo-Confucians also reconstructed Confucian tradition under the impact of Buddhism but saw it as a return to the Confucian *Analects*. The pioneers of both Vedanta and neo-Confucianism were, in fact, charged by critics of being crypto-Buddhists—in the way Aquinas was charged with exhibiting Averroist tendencies. Nevertheless, in every one of these cases there is perceived a need to absorb an external impact without appearing to do so, and a return to the origins myth allows such a process to be carried through without excessive exposure to possible obstacles from religious or cultural conservatives.[11]

This does not mean that there are no grounds for making such reinterpretive claims possible. As we saw earlier, many traditions have sufficient resources for rerooting major themes absorbed from the outside as articulations of minor themes within the tradition. However, there is also another factor that, in the European case, facilitated the absorption of Arabic traditions of science and philosophy as simply a return to the Greek heritage, and treating the Arabic culture as a mere carrier of ancient European thought. Outside the domain of religious revelation, Arabic thinkers were unusually generous in acknowledging what they received from other cultures—even if in the process the ideas imported became radically transformed. In doing so they saw themselves as obeying the Quranic dictum to "acquire knowledge, even if it be from China." The pioneer of Arabic philosophy, al-Kindi, wrote:

> We owe great thanks to those who have imparted to us even a small measure of truth, let alone those who have taught us more, since they have given us a share in the fruits of their reflection and simplified the complex questions bearing on the nature of reality. If they had not provided us with those premises that pave the way to truth, we would have been unable, despite our assiduous lifelong investigations, to find those true primary principles from which the conclusions of our obscure inquiries have resulted, and which have taken generation upon generation to come to light heretofore.[12]

In support of his view, al-Kindi quotes Aristotle:

> We ought to be grateful to the progenitors of those who have imparted to us a measure of truth, just as we are to the latter, in so far as they have been the causes of their being, and consequently of our discovery of the truth.[13]

Nevertheless, given the political exigencies of the time, the European inheritors of Arabic science, philosophy, and culture were in no position to emphasize this influence. Instead they invented the doctrine of a dark age that intervened between New Europe (the West) and Old Europe (the Hellenistic world) to facilitate the assimilation of Arabic scholarship into medieval Europe. This doctrine is surely one of the strangest historical notions ever invented. In the history of cultural evolution it has no parallels. It requires us to see a remarkable discontinuity in European history not found elsewhere—in the Indian, Chinese, Egyptian, Mayan, or Aztec civilizations. Nowhere else do we encounter the kind of death and resurrection it imputes to European civilization.

What is even more peculiar is that this doctrine requires the civilization that declined in Europe in the past to have been preserved by another alien civilization, the Arabic, for nearly half a millennium, before being returned to its rightful heirs. These Arabic carriers of European civilization are themselves considered to belong to a separate high civilization who acted, so to speak, as keepers for a time, of the heritage of ancient Europe only to yield it back to its rightful progeny. Such an episode of unprecedented caretaking by one civilization of the heritage of another is surely unique in history. Equally unique is the European claim, made by no other civilization, that its own tradition became lost, was preserved by another alien civilization, and then returned to it as its rightful owner after a long time.[14]

Even more curious, the Dark Ages theory requires us to believe that the ancestors of medieval European civilization had rejected Hellenistic civilization as alien and pagan nearly 600 years earlier. Hence, that which is deemed legitimate inheritance by European Renaissance thinkers had been pushed out by their predecessors as an alien culture. It was this earlier repudiation that had ensured that much of Hellenistic science and philosophy did not become a part of Western Europe after the fall of Rome.[15]

Equally odd is the location for the resurrection of the lost heritage. It was not resurrected in its original geographical location in Greece or in a part of the original Hellenic territory, but far away from its Mediterranean heartland by people who had destroyed the ancient civilization by entering Europe as nomadic invaders. By contrast the Arabic, so-called "mere" carriers of the Hellenic heritage, were themselves located in precisely those regions that included most of the original Hellenic geographical arena. Indeed Hellenic culture, after the conquests of Alexander, extended its influence from Persia to the lands of the eastern Mediterranean, Egypt, parts of Italy, and Greece. Thus, the Arabic civilization, which is supposed to have been merely a transmitter of Hellenic culture, actually embraced the greater part, as well as the more Hellenized parts, of Hellenistic civilization (with the exception of Greece). Yet, in spite of this geographical continuity, the

Renaissance model denies that Arabic civilization is also a rightful heir of Hellenic culture.

Moreover, the ancient Greeks saw their cultural world—and what they considered to be its legitimate heirs—differently from the modern Europeans after the Renaissance era. When Alexander of Macedonia attempted to unite the world, he conquered territories to the east in Persia and south in Egypt in order to accomplish his mission. His conception of uniting the world under Hellenism did not include the geographical area of medieval Europe or perceive ancient Greece as rightly belonging to the European north. Moreover, it is likely that such identification would have been rejected by the Greeks, who generally saw their civilization as historically influenced by the cultures of Mesopotamia and Egypt.

Even worse, the Roman Empire, which in many ways was the real predecessor of medieval European culture since it embraced those territories that later became its core areas—an empire that Charlemagne attempted to restore as the Holy Roman Empire in 800 CE—never took Hellenic civilization very seriously (apart from some elements of it such as Stoic philosophy). Actually the rule of Rome saw the centers of Hellenic learning shift to the eastern and southern portions of the original Hellenistic culture. Indeed in areas under Rome's dominion the Greek tradition of learning became increasingly emasculated; while the Romans showed an interest in letters, they had hardly any deep involvement with Greek philosophy, even less with its science.

In this regard nothing could be more revealing than the contrast between the Roman and Arabic civilizations when both came in contact with the Hellenic heritage. Although the Romans had made contact with the Greek tradition many centuries earlier, they hardly took it any further—though they did accord it some respect. It was the Arabic thinkers who embraced Greek philosophy and science with enthusiasm as their legitimate inheritance. This embrace even created for them very serious intellectual conundrums concerning the reconciliation of Greek philosophy and science with the revelations of their faith. These came to constitute some of the central issues in Arabic philosophy, and the tradition of thinking they developed profoundly influenced European scholastics after they came in contact with it.

Perhaps the most surprising feature of the Dark Ages model, and its view of Arabic civilization as a mere carrier and European civilization as the legitimate heir of the Greek tradition, is that only within the Arabic civilization did Greek thought undergo sustained development. Throughout the heyday of Arabic civilization, Aristotle was regarded as "The Philosopher." By contrast Aristotle became a part of the European heritage only in the thirteenth century with the scholastics, but was quickly rejected in the sixteenth century when Europe began to develop new ideas in opposition to the ancients. Significantly, Aristotle was not rejected by the Arabic world until after al-Ghazali in the eleventh century and, in spite of that, has continued to live on until today as a major influence through the philosophies of al-Farabi and Ibn Sina (Avicenna). What appears strange about the view of Arabic culture as carrier, and European culture as rightful heir, of Hellenism is that only Arabic civilization has taken the Greek tradition seriously enough to make it a part of its living tradition. By contrast the Europeans rejected it as pagan and alien in the fifth century, absorbed it as their rightful heritage only for a brief period in the Middle Ages, and then subverted it shortly thereafter! It is difficult to conclude that Greek thought has ever been taken seriously in Europe (excluding the brief interlude of the medieval period beginning in the twelfth century) except as a tradition historically linked to Western thought.

This raises the suspicion that the narrative of a dark age in Europe, with its mythology of heritage lost and found, was only created in order to avoid the charge—a highly incendiary one given the struggle of Europe against Arab-Muslim colonization in the medieval era—that the translations of Arabic texts involved the transmission of Arabic philosophy and science into Europe. The subsequent political and economic domination of the world by Europeans only served to consolidate this myth and entrench it more deeply.

Hence it appears reasonable to conclude that the account of the Renaissance in Europe as the rebirth on European soil of its legitimate ancestral Greek heritage is unsustainable. Moreover, given the large impact of Arabic culture and Chinese technology in shaping events in Europe at that time this view, even if taken to be true, can only partially explain the phenomenon of the European Renaissance. Indeed it is probably more accurate to speak of the Renaissance as the time when Arabic scientific and philosophical traditions from Iberian territories in Western Europe under Muslim rule, and Chinese technologies from those parts of Eastern Europe under Mongol rule, met in medieval Europe sandwiched at the center. It is this potent brew, and the dialogue it initiated within the European world, that provided a major context for both the Renaissance and the birth of modern science in Europe.

Chapter 7

The Narrow Copernican Revolution

Let us now examine more closely the specific problem we are addressing. Is modern science simply the outcome of ideas and problems rooted in the European, especially Greek, tradition of philosophy and science alone, or is it the result of the meeting of ideas from many different cultures, albeit within Europe? One of the main reasons for considering modern science to be the direct descendent of the Greek tradition is the central status assigned to the Copernican heliocentric theory in bringing about the birth of modern science. According to historian of science Thomas Kuhn the changes that took place in the seventeenth century, culminating in the Newtonian synthesis of both terrestrial physics and celestial astronomy, can be interpreted as unintended consequences of the Copernican hypothesis. These consequences followed because the moderns, unlike the ancient Greeks, refused to view planetary theory merely as concerned with building mathematical models designed, in Ptolemy's words, "to save the phenomena." Instead Galileo and Kepler, who pioneered the way to Newton, treated the Copernican theory as a physical hypothesis that described actual states of affairs in the heavens. This approach led Galileo to create a new kinematics that could better accord with the Copernican model and led Kepler to discard the notion that all heavenly motions had to be explained in terms of circular motions. Instead Kepler replaced the "dogma of the circles" for planetary motions, instituted by Plato, with the new idea that planets performed elliptical orbits around the sun. Finally Newton united these two seminal achievements by proposing the concept of gravitation that allowed him, in conjunction with his three laws of motion, to deduce the kinematical and planetary laws discovered by Galileo and Kepler. Kuhn holds that Newton's synthesis can be viewed as culminating the process of intellectual transformation initiated by Copernicus that began the new science of the modern era. Although these historical changes took place over a period of 140 years, beginning with Copernicus' publication of the *Revolutionibus* in 1543 and ending with Newton's *Principia* in 1684, Kuhn considers the final result to have been, in a sense, already inherent in the Copernican theory.[1]

Yet the original work published by Copernicus includes an introduction designed to warn the reader not to take the heliocentric theory as a physical model. It stresses the point that the new theory is only offered as a mathematical hypothesis—albeit one different from Ptolemy's hypothesis:

> Since the newness of the hypothesis of this work—which sets the earth in motion and puts an immovable sun at the center of the universe—has already received a great deal of publicity, I have no doubt that certain of the savants have taken grave offense and think it wrong to raise any disturbance among liberal disciplines which have had the right set-up for a long time now. If, however, they are willing to weigh the matter scrupulously, they will find that the author of this work has done nothing which merits blame. For it is the job of the astronomer to use painstaking and skilled observation in gathering together the history of the celestial movements, and then—since he cannot by any line of reasoning reach the true causes of these movements—to think up or construct whatever causes or hypotheses he pleases such that, by the assumption of these causes, those same movements can be calculated from the principles of geometry for the past and for the future too. (Copernicus, trans. Wallis 1995, p. 3)

Some historians have argued that the introduction in the *Revolutionibus*, originally believed to have been the work of Copernicus, was really penned by Andreas Osiander, a Lutheran theologian and friend of Copernicus, who finally saw the study through the press (Kuhn 1957, p. 187). Such a conclusion appears warranted since in the text of his work Copernicus addresses objections to his theory that he would have taken into account only if he considered his theory to give a physically real description of the universe. For example, he addresses the argument that clouds and other things floating in the air, or things falling toward or rising away from the ground, would be left behind by a rotating earth. He counters them by proposing that the perpetual rotation of the earth causes the atmosphere itself—or that part of it close to the earth—to acquire the same motion and participate in it on account of its contiguity. Hence, he concludes that "the air which is nearest the earth and the things floating in it will appear tranquil, unless they are driven to and fro by the wind or some other force" (Copernicus, p. 17). These are not serious charges if Copernicus took his theory to be only a mathematical hypothesis designed to serve as a better computing tool for describing and predicting heavenly motions—one that did not require us to presume that the earth was, in fact, rotating.

Nevertheless, even if Kuhn rightly perceives that the assumption of physical realism with regard to the Copernican hypothesis played a crucial role in triggering the Scientific Revolution, the question remains as to why this ontological assumption came to be entertained seriously. Why did it catch on? More specifically why did it catch on in spite of the many obvious objections to a physical realist interpretation of the theory—objections presumably addressed in the introduction to the *Revolutionibus*, which disclaimed any physical realist interpretation of the theory? Moreover, the heliocentric theory had been known to the ancient Greeks along with the reasons for its physical implausibility. Copernicus himself refers to the Pythagoreans Herakleides and Ekphantus, and Hicetas the Syracusean, who believed that the earth revolved at the center of the universe (Copernicus, p. 13). He was also aware that Aristarchus had entertained the possibility that the earth revolved around the sun, and that this theory had been rejected by no less an authority than Aristotle largely on physical, rather than mathematical, grounds.

The obstacles to making the earth move were also known to Indian astronomers. Alberuni's study of Indian science and thought in the eleventh century makes this point:

> As regards the resting of the earth, one of the elementary problems of astronomy, which offers many and great difficulties, this too, is a dogma with the Hindu astronomers. Brahmagupta says in the *Brahma Siddhanta*: "Some people maintain that the first motion (from east to west) does not lie in the meridian but belongs to the earth." But Varahamihira refutes them by saying: "If that were the case, a bird would not return to its nest as soon as it had flown away from it towards the west." (Alberuni 1983, vol. 1, p. 276)

He also writes:

> The followers of Aryabhata maintain that the earth is moving and heaven resting. People have tried to refute them by saying that if such were the case, stones and trees would fall from the earth. But Brahmagupta does not agree with them, and says that would not necessarily follow from their theory apparently because he thought that all heavy things are attracted towards the center of the earth. (Alberuni 1983, vol. 1, pp. 276–277)

Clearly the Indian astronomers from the sixth century onward, following Aryabhata, had also considered the problem that a rotating earth would leave objects behind, and even developed auxiliary theories to explain away this possibility. Copernicus seems to think this would not occur because a rotating earth would impart its motion to the atmosphere, which presumably would carry other bodies in it with it. By contrast the Indian astronomer Brahmagupta seems to have proposed a theory of gravitation to resolve the problem according to the aforementioned passage from Alberuni. Although these answers are different from the concept of inertia used today to deal with the same objections, they nevertheless reveal that the question of whether a moving earth theory can be reconciled with physical realism confronted astronomers in many cultures. Alberuni himself adds that assuming that the earth rotates does in no way impair the value of astronomy, as all appearances of an astronomical character can be explained quite as well as when we assume the heavens rotate (Alberuni 1983, vol. 1, p. 277). Nevertheless, he concludes that there are other reasons (which he does not give, presumably because he assumed that Arabic astronomers would be aware of the standard physical objections already raised by Aristotle) that make rotation impossible.

Since the possibility of the motion of the earth—as an astronomical and mathematical model—had been entertained in a number of cultures (the Hellenic, the Indian, and the Arabic), and rejected by most astronomers on physical grounds, we are forced to ask why Galileo and Kepler came to take it seriously? One answer is provided by Cohen, who argues that the realist core and mathematical harmony of the Copernican hypothesis exerted such a spell over Galileo and Kepler that "they were inspired to overcome all objections arising from commonsense, from everyday experience, from God's word taken literally, and from elementary Aristotelian doctrines on local motion" (Cohen 1994, p. 510). However, Cohen's answer does *not* provide an explanation; it only expresses wonderment. It tells us how impressed Galileo and Kepler were by Copernican realism, but nothing about the reasons that impressed them in the first place.

In finding an answer to this question we will come to one of those key issues that will decide whether we can consider modern science—the science that defined itself with Newton's synthesis—to be a continuation of the Hellenistic tradition considered to be already on its threshold, or to constitute a distinct break with the past. The continuity thesis assumes that immediately after Ptolemy (and without a lapse of nearly a millennium and a half) there could have been some Greek Copernicus who could have proposed and constructed a heliocentric model. Moreover, by taking this theory as physically real, Greek science could have directly led to the theory proposed much later by Newton. That it did not actually happen is beside the point—it could have happened because Greek science had all the relevant mathematical, philosophical, and conceptual tools to make this transition. It only needed a heliocentric realist to transpose the positions of the earth-moon with the sun, to trigger the other changes needed to accommodate this hypothesis, and to force the slide more or less inevitably in the direction of modern science. This is the reason why Hellenic science can be rightly seen as the direct precursor of modern science.

Apart from Kuhn another historian who has argued along these lines is Dijksterhuis. He contends that the arguments advanced in Book I of *De Revolutionibus* drew only upon ancient sources to defend the heliocentric theory as physically true, and that Books II through VI followed Ptolemy's *Almagest* closely, deploying the instruments of epicycles and eccentrics but, significantly, without Ptolemy's equant. Therefore, concludes Dijksterhuis, "barring the application of trigonometric methods of computation one finds nothing in [*De Revolutionibus*] that might not just as well have been written in the second century CE by a successor of Ptolemy."[2] Hence, it is easily conceivable that the seventeenth-century Scientific Revolution in Europe could have taken place much earlier in the Hellenistic world.

If Dijksterhuis is right, then there is hardly any leeway for a dialogical interpretation of the Scientific Revolution as the outcome of influences from a plurality of cultures. Even if it could be shown that multicultural influences did play a role, they would have been redundant since the revolution could have taken place without them. Nevertheless, Dijksterhuis does qualify his conclusion by saying that the novelties in Copernicus compared to those in Ptolemy involved the application of trigonometry and the rejection of the use of the equant in astronomy. What he fails to see is that Copernicus inherited his trigonometric methods from Indian mathematical astronomers, and rejected the equant on grounds largely motivated by the critiques of Ptolemaic astronomy by Arabic astronomers. Examined more systematically, these influences are not as insignificant as they might appear at first sight—indeed they are key elements without which there would have been no modern scientific revolution.

Moreover, Dijksterhuis has also overlooked the Indian place-value number system with zero. Without this number system Copernicus would have found it more difficult to manage the complex computations he was required to make in order to reread the observational data he used, obtained from a moving earth, from the point of view of the stationary sun. Laplace in 1814 described the significance of the system thus:

> The ingenious method of expressing every possible number using a set of ten symbols (each symbol having a place value and an absolute value) emerged in India. The idea seems to be simple nowadays that its significance and profound importance is no longer appreciated. Its simplicity lies in the way it facilitated calculation and placed arithmetic foremost amongst useful inventions. The importance of this invention is more readily appreciated when one considers that it was beyond the two greatest men of Antiquity, Archimedes and Apollonius. (Ifrah 2000, p. 361; see also Dantzig 1954, p. 26)

However, even if Copernicus may have done without the Indian number system, it is certain that the wider Copernican Revolution associated with the names of Galileo, Kepler, Descartes, and Newton would not have come to fruition, and Copernicus is likely to have remained as minor a figure in science as Aristarchus who preceded him.

There is also good reason to suppose that the Indian mathematical contributions may have been greater—especially if we take into account the mathematical innovations that made possible the wider Scientific Revolution culminating with Newton. These innovations include infinite series representations of trigonometric and circular functions, and early notions of differentiation and integration in modern calculus. Indian mathematics may have made crucial contributions to these discoveries. If so, we cannot adequately acknowledge them by way of summary statements referring to the invention of zero, the decimal place system, and trigonometry. Such brief and perfunctory descriptions serve to mask rather than illumine the role that Indian mathematics played in the genesis of modern science.

In order to fully appreciate the contributions of Indian mathematics we have to understand the historical context within which it developed. Many of its achievements took place only after Greek astronomical models, and their associated techniques, were introduced into India, and after Indian mathematical astronomers increasingly switched to these models after appreciating their superior predictive accuracy. According to Pingree, at least four different Greek texts on astronomy were transmitted to India in the second, third, and fourth centuries. Indian texts based on these were composed in Sanskrit from the second to the seventh centuries—so much so that it is possible to recover early Greek non-Ptolemaic astronomical texts by studying Indian Sanskrit texts (Pingree 1976).

However, the Indian mathematical astronomers did not passively receive the Greek texts. They used them as the basis for developing new and more powerful mathematical techniques that went far beyond Greek computational and algebraic achievements. According to Joseph, the earliest, and one of the most important, of the Indian mathematical astronomers was Aryabhata (476–550 CE). His book *Aryabhatiya*, written in 499 CE, summarizes many of the achievements of Indian mathematical astronomers before him and became a canonical text in Indian astronomy. It gave details of an alphabet-numeral system and rules for arithmetical operations that greatly enhanced computational facility, and anticipated most of the features of the current number system we use. He also gave general rules for summing natural number series and their squares and cubes. One of the major innovations in his pioneering work is the introduction of the sine and versine functions for the first time in history, which laid the basis for modern trigonometry (Joseph 2000, pp. 265–267). These pioneering paths that opened up in the *Aryabhatiya* were to be extended and combined over the next millennium, culminating in the achievements of the Kerala School.[3]

In the seventh century the mathematical astronomer Brahmagupta (598–670 CE), in his astronomical treatise *Khanda Khadyaka,* advanced trigonometric computation techniques by describing a method for obtaining the sines of intermediate angles from a given table of sines. His achievement is equivalent to the Newton–Sterling interpolation formula up to second-order differences known to European mathematicians only many centuries later. The Arabic translation of his work, *Brahma Sputa Siddhanta,* became the route through which Indian astronomy, along with all the powerful mathematical techniques associated with it, entered the Arabic world and the West (Joseph 2000, p. 267).[4]

The last Indian mathematical-astronomer to have a direct impact on Arabic scholars was Bhaskara II (1114–1185 CE). Bhaskara's *Bijaganita* deals with many algebraic problems, including evaluating surds and solving simple and quadratic equations. He also provides general rules for solutions of indeterminate equations. His cyclic method of solving indeterminate equations of the form $ax^2 + bx^2 + c = y$ is now associated in the West with the name of William Brouncker, who discovered it in 1657. Moreover, in his *Siddhanta Siromani* Bhaskara deals not only with relations between different trigonometric functions but also with certain preliminary concepts of the differential calculus and analysis. For example, in computing the instantaneous motion of a planet Bhaskara uses the notion of an "infinitesimal" unit of time between its successive positions. The value he gives for this, 1/33,750 second, is extremely small. The procedure he adopts is analogous to the use of "indivisibles" by Cavalieri, Roberval, and Fermat, who pioneered the discovery of calculus in Europe before Newton and Leibniz. Its use shows that Bhaskara recognized the concept of a differential. Bhaskara was also aware that when a variable attains its maximum value the differential becomes zero. Furthermore he recognized that when a planet is farthest from the earth or closest to it, the equation of the center vanishes, and he inferred that the differential of the equation of the center is zero for some intermediate position. These ideas anticipate important discoveries associated with the modern calculus in the seventeenth century in Europe (Joseph 2000, pp. 299–300).

However, we cannot confine ourselves to only examining that part of Indian mathematics transmitted to the West through astronomical treatises translated and available in Arabic. This may have been acceptable in the past, when it was assumed that no significant Indian mathematical contributions reached the West after Bhaskara for two reasons. First, it was thought that no further significant advances took place in Indian mathematics after Bhaskara's work. Second, it was assumed that no new Indian works could reach the West after the fourteenth century because Arabic translations of Indian works ceased as Arabic civilization itself turned inward after the

Mongol invasions that destroyed the Abbasid Caliphate. We now have good grounds for supposing both assumptions are false. First, from the fourteenth to sixteenth centuries there was a revolution in Indian mathematics as the Kerala School developed new methods that used infinite series representations of trigonometric functions, and discovered techniques that anticipated many elements of modern calculus. Second, there is evidence that these discoveries were probably directly communicated to Europe after the Portuguese found a trade route to India in 1498. This suggests that we cannot ignore the possibility that the Kerala School of Indian mathematics influenced the Scientific Revolution in modern Europe. (Joseph 2000, p. 356)

Let us begin by examining the achievements of the Kerala School before looking at its possible influence on European mathematicians. According to Joseph, problems in astronomy, which had been the main motivations behind the Indian mathematical developments since the time of Aryabhata, continued to motivate research in the Kerala School. Since astronomical work required highly detailed trigonometric tables, Kerala mathematicians were led to discoveries that were repeated by Europeans much later. For obtaining highly accurate estimates of π, and for calculating rational approximations for different trigonometric functions, they were led to develop infinite series representations of their values. Although the Kerala School itself is described as a chain of mathematicians whose work extended from the fourteenth to the eighteenth centuries, we will focus on the accomplishments of three of them—Madhava (1350–1425 CE), the pioneer of the Kerala School, who made the decisive step of going beyond the finite procedures of ancient mathematics to deal with infinite series; his follower Nilakantha (1444–1544 CE), who not only extended Madhava's discoveries, but also instituted a revolution in astronomical theory that possibly influenced modern astronomy; and Jyesthadeva (1500–1575 CE), who wrote the unique study *Yuktibhasa* that brought together the achievements of the Kerala School over the previous two centuries, and possibly laid the basis for its discoveries to enter modern science.

Madhava was the greatest of the Indian medieval mathematical astronomers. According to Rajagopal and Rangachari he "took the decisive step onwards from the finite procedures of "ancient" mathematics to treat their limit-passage to infinity which is the kernel of modern classical analysis" (Rajagopal and Rangachari 1978, p. 101). To him is attributed the discovery of the power series for inverse tangent now attributed to Gregory; the power series for π associated with Leibniz; the power series for sine and cosine named after Newton; and approximations for sine and cosine functions to the second order of small quantities linked with Taylor (Joseph 2000, p. 289).

The major work of Nilakantha is the *Tantrasangraha*, an astronomical treatise written in 1501, which uses epicycles and eccentrics in its mathematical model and treats of ways to compute planetary positions by applying techniques developed by the Kerala School. His work is important in two respects. First, as a mathematical treatise, it extends the discoveries of Madhava in infinite series by discovering new series and techniques that facilitated more rapid convergence. Second, as an astronomical treatise, it proposed a geoheliocentric model of the planetary system that influenced future Indian astronomical studies—and possibly the Scientific Revolution in Europe.

Jyesthadeva is important mainly because of the summary he provided of the discoveries of the Kerala School, which made it possible for its achievements to be more widely disseminated. In it we find a formulation of the famous Madhava infinite series for inverse tangent that is often attributed to Gregory (1638–1675), a pioneer of infinite series studies in the modern era, expressed as follows:

> The first term is the product of the given Sine and radius of the desired arc divided by the Cosine of the arc. The succeeding terms are obtained by a process of iteration when the first term is repeatedly multiplied by the square of the Sine and divided by the square of the Cosine. All the terms are then divided by the odd numbers 1, 3, 5, …. The arc is obtained by adding and subtracting [respectively] the terms of odd rank and those of even rank. (Joseph 2000, p. 290)

The above formulation of the rule for generating the infinite series for inverse tangent is given in terms of the Indian sine and cosine expressed in capital letters such that $\text{Sin}\,\theta = r\sin\theta$ and $\text{Cos}\,\theta = r\cos\theta$, where r is the radius of the arc. Using modern notation, Madhava's rule may be written as follows:

$$r\theta = \frac{r\text{Sin}\,\theta}{\text{Cos}\,\theta} - \frac{r(\text{Sin}\,\theta)^3}{3(\text{Cos}\,\theta)^3} + \frac{r(\text{Sin}\,\theta)^5}{5(\text{Cos}\,\theta)^5} - \cdots$$

This rule becomes the power series for inverse tangent when we write it using the modern concepts of sine and cosine:

$$r\theta = \frac{r(r\sin\theta)}{1(r\cos\theta)} - \frac{r(r\sin\theta)^3}{3(r\cos\theta)^3} + \frac{r(r\sin\theta)^5}{5(r\cos\theta)^5} - \cdots$$

Therefore:

$$\theta = \tan\theta - \frac{\tan^3\theta}{3} + \frac{\tan^5\theta}{5} - \cdots \; [\text{Madhava Series for Arctan}]$$

The above formula is equivalent to the Gregory series for inverse tangent.

These accomplishments of the Indian mathematicians are directly relevant to the central question we set out to answer: Was Greek science on the threshold of modern science because it was only one step away from the Copernican Revolution? The answer depends on the significance we assign to the discoveries of Indian mathematicians, not only in making possible the work of Copernicus but also in facilitating the changes that led to its culmination in the Newtonian synthesis. We can separate our approach to an answer into two parts. First, we can consider the significance of the earlier part of Indian mathematics transmitted through the Arabic tradition. Second, we can examine the significance of the contributions of the Kerala School. We have already provided the answer for the earlier discoveries—the Copernican Revolution could not have consolidated itself through the wider Scientific Revolution without the Indian mathematical discoveries in trigonometry and the number system.

However, the answer to the role of mathematical contributions from the Kerala School would depend on whether we consider early modern developments in European mathematics in infinite series and calculus to have been created independently of similar Indian achievements, or to have been influenced by them. Indeed if we adopt the thematic transmission criterion proposed earlier we would be compelled to conclude that the Kerala School influenced modern European mathematics.[5]

First, the arrival of Vasco da Gama in 1498 at the port of Calicut, which happened to be located in the center of mathematical developments in Kerala, in the lifetime of Nilakantha and shortly ahead of the birth of Jyesthadeva, opened a corridor of communication for the transmission of the Kerala School's discoveries to Europe. Second, the Portuguese not only were motivated to exploit such opportunities for learning from Indian mathematical astronomers, but also had trained personnel who were able to do this. An example is Matteo Ricci (1552–1610), who was later to acquire fame as a leading Jesuit astronomer in the service of the emperor in China. He had first stayed in India for a few years before leaving for China, and continued to remain in close contact with the rector of the *Collegium Romano*, the primary institution for educating Jesuits (Joseph 2000, p. 356). Finally, dominant themes in Indian mathematics, such as work on trigonometric and other series, as well as ideas of calculus, shortly afterwards became major themes in European mathematics—themes so new that they were considered to break with the classical tradition of mathematics. Adopting the thematic transmission criterion we would be prepared to say that not only did Indian mathematics reach Europe through Arabic scholars, but so did developments that had taken place after the collapse of the Abbasid Caliphate following direct contact with India.

There is also evidence that scientific developments in Europe were influenced directly by thematic ideas in the *Yuktibhasa* of Jyesthadeva. Written in 1550, a little over half a century following the arrival of the Portuguese in India, and at the time the Portuguese were actively setting up numerous missionary schools in the area where Jyesthadeva lived, the *Yuktibhasa* summarizes all the significant achievements of the Kerala School over the previous two centuries including the discoveries of Madhava and Nilakantha. It appears to have been a work written to influence and inform Indian students of European astronomy taught in the missionary schools, and to show them that the computational astronomical discoveries of the Kerala School should not be underestimated. This would explain the changes in the way these achievements are presented in the *Yuktibhasa*. It is a unique text for the time within the Indian context for two reasons. First, it ruptures with tradition because it is written in Malayalam, the regional language of Kerala, and not Sanskrit, the traditional language used by Jyesthadeva's predecessors and the intellectual elites where he lived. Second, it again breaks with tradition by giving detailed demonstrations of theorems and derivations of the rules it describes, instead of merely presenting them without proof.

This dramatic reorientation of language and mode of argument in the *Yuktibhasa* suggests that it is not written for a traditional audience versed in Sanskrit and one likely to accept the results it describes as given. It appears intended for an audience interested in the discoveries of the Kerala School but not conversant with Sanskrit—otherwise Jyesthadeva is unlikely to have made the effort to translate ideas from Sanskrit to the regional language. Hence its intended audience cannot have been the traditional intellectual elites who would have been proficient in Sanskrit. Moreover, since the audience required proofs it cannot have been intended for Indian technicians who would only have wanted to use the rules it describes to make calendars, draw navigations charts, or cast horoscopes.

It is reasonable to conclude that the audience could only have been the students being trained in the local language Malayalam at the numerous new missionary schools set up by Jesuits in India. Their education in the local language would also have included training in the European astronomical and mathematical tradition, because one of the main reasons Jesuits were welcomed by rulers in the East was their superior skill in mathematical and astronomical computations. It also helped that being taken seriously as mathematical astronomers gave Jesuits the freedom to move around and conduct their missionary activities—a discovery that served them well when they later entered China. Students in the Jesuit schools would not have been impressed by the discoveries of the Kerala School if they had been handed down as inspired by authority. Neither would they have understood them in Sanskrit. In the past the Indian mathematicians may have taught the proofs to their disciples, but not included them in the texts that disseminated their discoveries to craftsmen. However, it is likely that in order to impress upon these students the achievements of the Kerala School, Jyesthadeva found it necessary to write in the vernacular language and include proofs.

However, there are also thematic changes in the European tradition that parallel major themes in the *Yuktibhasa* and that developed more than half a century after its publication in India. This suggests that the book intended for Indian students may also have later influenced European developments. One important reason for suspecting such influence is that European mathematicians shortly afterward began to approach problems associated with infinite sums using the method of direct rectification described in the *Yuktibhasa*. This method went beyond the method of exhaustion known to the ancient Greeks, including Archimedes. The method of exhaustion involves calculating the area of a shape, or the length of a curved line, by inscribing a sequence of polygons whose areas, or circumference, converge to that of the shape concerned. Archimedes had used the method of exhaustion to compute the area and circumference of a circle by filling the circle with polygons of increasing number of sides. As the number of sides tend to infinity the area, or circumference, of the polygons tend to the area, or circumference, of the circle.

The influence of Archimedes on European mathematicians can be traced to the translation of his works in 1543, about seven years before Jyesthadeva wrote the *Yuktibhasa*. Hence, it is possible that his work may have played a role in making Europeans receptive to the more powerful techniques in the *Yuktibhasa*. Joseph describes the difference between the Archimedean method of exhaustion and the Indian approach by using the example of the derivation of the arctan Madhava series given in the *Yuktibhasa*:

> The approach involves what is known as the direct rectification of an arc of a circle, i.e. the summation of very small arc segments and reducing the resulting sum to an integral ... This is a very interesting geometric technique different from the method of exhaustion used in the Arab and European mathematics. In the Kerala case you are sub-dividing an arc into unequal parts and while in the other (Arab and European) case there is a sub-division of the arc into equal parts. The different technique used in Kerala was not because the method of exhaustion was unknown to the Indians. Indeed, it is likely that Aryabhata used the method of exhaustion to arrive at his accurate estimate of the circumference for a given diameter. The "exhaustion" method was probably avoided because the calculation involved working out the square roots of numbers at each stage of the calculation, a tedious and time-consuming task. (Joseph 2000, p. 131)

The Indian method of summing mathematical series and computing areas, using the method of rectification, probably developed from a geometrical approach to representing number series. The technique involves breaking the area of a geometrical figure into segments in such a way that each segment of the figure has an area represented by a term of the number series. The sum of the series can then be computed to give the area of the figure. Nilakantha adopts this method in solving problems in arithmetical progressions (Mallayya 2000); Jyesthadeva uses it to derive the Madhava–Gregory inverse tangent series in the *Yuktibhasa*. This derivation not only shows how trigonometric functions can be expressed in the form of an infinite series but, according to Ramakrishnan who examined the derivation in detail, it also "involves several techniques including the idea of integration and differentiation"[6] (Ramakrishnan 2002, p. 138).

A second thematic parallel between the *Yuktibhasa* and developments in European science after its publication is that the planetary theory it adopts as a mathematical model for computation purposes, and which was first proposed by Nilakantha, is similar to that proposed a few decades later by Tycho Brahe as an alternative to the Copernican theory. It raises the possibility that Brahe may have been influenced by the discoveries of the Kerala School described in the *Yuktibhasa,* and constitutes another dominant theme of the Kerala School that became a theme of European science in the decades following European contact with India.

In order to understand Nilakantha's motivation in proposing his new theory we have to consider closely the methods adopted by Indian mathematical astronomers before him when they set about computing planetary positions, that is, planetary geocentric longitudes. They adopted two different strategies: one for the inner planets, Mercury and Venus, which had long been known to keep close to the sun, and another for the outer planets, Mars, Jupiter, and Saturn, which strayed far from it. Both approaches involved computations that were made in a three-step procedure set down nearly a millennium earlier by Aryabhata (Sriram 2002, p. 105).

Let us first consider the method used to compute the geocentric longitude of the outer planets. The first step involved calculating the mean longitude of the planet. For the purpose of this calculation Indian astronomers assumed that the planet moved at uniform speed in a circular orbit around the earth from its position at the beginning of the current epoch which they dated back to February 17, 3102 BCE—what in the Indian tradition is seen as the beginning of the *Kaliyuga* (Age of Kali). To find the current position they computed the number of civil days (called the *ahargana)* that had elapsed since the beginning of the current epoch. By adopting this day-counting system Indian astronomers evaded problems associated with religious calendars where the conventional dating could begin with different years. The resultant position computed can be described as the mean planet—an imaginary position based on the assumption of where the planet would have been had it traveled with its mean motion from the origin of the epoch.

However, it is not the true position of the planet. The reasons for this are well known today. First, the planets move in heliocentric orbits that are elliptical and their paths do not always coincide with the ecliptic, that is, the apparent path of the sun traced against the stars as a result of the earth's orbit around it. Second, the speed of the planet also varies according to its distance from the sun (Girish and Nair 2002, p. 84).

However, the Indian astronomers saw these nonuniform motions as a result of the attraction of the planets to points in the heavens which they named the *mandocca* and the *sighrocca* of the planet. These points were seen as generating an attractive force on the planet that caused it to deviate from its uniform circular motion. The *mandocca* is comparable to the modern apogee of a planet when it moves most slowly, and can be determined by finding the point of slowest angular motion of the planet as seen from the earth. The *sighrocca* for the outer planets was the sun. Indian astronomers saw the forces from the two attractive centers as causing the planet to deviate from its computed mean path so that two corrections— the *manda* and *sighra* corrections—had to be made in order to determine the true position of the planet.

Expressed in modern terms the *manda* correction can be seen as accounting for the eccentricity of the planet's orbit and the *sighra* correction as accounting for the retrograde motion, and other distortions, that result from the relative motion of the planet and the earth as they both revolve around the sun. Making both these corrections on the mean planet yields the true geocentric longitude of the planet (Girish and Nair 2002, p. 86–89).

However, the above procedure only describes how Indian mathematical astronomers approached computations dealing with outer planets, at least prior to Nilakantha. By contrast when it came to the inner planets, Mercury and Venus, they adopted another approach. Since these planets did not stray far from the sun, they took the mean planet as the mean sun. The *manda* correction for the planet was applied on the mean sun. Moreover, the *sighrocca* center of attraction of these planets was not taken to be the sun. Instead it was seen as located at the mean heliocentric longitude of the inner planet concerned (Sriram 2002, pp. 106–107). The reason for their different approach to the inner planets is not clear, but it is likely that, given the close attachment of the inner planets to the sun, it was difficult for Indian astronomers to conceive them as being drawn to another attractive center that led them far away from the sun (as was the case with the outer planets).

Nevertheless, it was known that their predictions for the inner planets were not as accurate as those for the outer planets, and prompted numerous attempts to improve astronomical predictions. In 1500 Nilakantha triggered a computational revolution in the Kerala School tradition by bringing the two approaches under the same method, and improving computational accuracy for the inner planets in the process. He did this by taking the mean planet—even for the inner planets—as the value upon which to apply the *manda* correction. He also assumed that the mean sun was the *sighrocca* for the inner planets, as it was for the outer ones. This led to much improved predictions for the positions of the inner planets, and unified the approach to planetary theory to such an extent that it came to be universally adopted by leading astronomers of the Kerala School (Sriram 2002, p. 112).

As a result of the changes he made Nilakantha shifted the Indian astronomical tradition to the beginning of the sixteenth century, and shortly after the arrival of Europeans in India, into effectively adopting a geo-heliocentric model for computing planetary positions that anticipated a similar shift by Brahe in Europe at the end of the century. The parallels between the Nilakantha and Brahe models have not gone unnoticed. In their paper, *Yuktibhasa of Jyesthadeva,* Sarma and Hariharan write:

> If the *sighra* is identified with the sun itself then this agrees broadly with the modern theory with the positions of earth and sun reversed. In fact western astronomer Tycho Brahe (1546–1601) appears to have adopted a similar theory. (1991, p. 193)

Joseph and Almeida also claim that Nilakantha's planetary model is similar to Brahe's model, and take it to be an important additional item of circumstantial evidence that the Kerala School influenced European scientists in early modern Europe (Aryabhata Group 2002, p. 44).

Even though Nilakantha and Brahe arrive at similar models for the planetary system, their motivations for making the change are not the same. Nilakantha proposed his theory in order to bring together two different computational approaches dealing with the inner and outer planets respectively; Brahe's model was intended to bring together two different theories, the heliocentric theory and Aristotelian physics, by adopting a model that made the planets revolve around a sun that revolved around the earth. But the thematic similarities between the Nilakantha model and the Brahe model give us one more reason to suspect that the influence of the Kerala School on modern European scientists might have been greater than hitherto known.

There is a third dominant thematic notion in the *Yuktibhasa* that also becomes an important theme of later European astronomy. This came about when European astronomers adopted the practice of using day numbers in the scientific specification of dates. It was introduced by Julius Scaliger in 1582, and is now known as the Julian day number system. Such a system was, as we have seen, adopted by Indian astronomers in order to eliminate ambiguities due to differences in calendar systems. The only major difference between the European and Indian systems is that the former begins its count from the biblical date of creation on January 1, 4713 BCE instead of from the start of the *Kaliyuga*. Joseph and Almeida argue that the new system, which parallels the Indian *ahargana* system, came to be adopted in Europe because its advantage would have been immediately evident to Jesuit astronomers in India. They take this to be one more item of circumstantial evidence that the influence of the Kerala School was felt in Europe (Aryabhata Group 2002, pp. 43–44).

Applying the thematic criterion again we can conclude that shortly after a new corridor of communication opened between Europe and Kerala (in India), and great interest was shown by Europeans in Indian calendar and computation techniques, many dominant thematic ideas of the Kerala School become dominant in Europe. This gives us reason to suppose that the discoveries of the Kerala School influenced the development of astronomical and mathematical ideas in early modern Europe. Moreover, there is also evidence that the *Yuktibhasa*, probably written only to educate Indian students in European missionary schools, had an influence that radiated into Europe. It influenced mathematical approaches, astronomical theories, and calendrical techniques in early modern Europe. If these arguments are accepted, then the European voyages to the East were more than voyages of geographical discovery—they were also voyages of intellectual discovery.[7]

We have seen that Dijksterhuis considered the use of trigonometry and the refusal to use the equant as the two differences that distinguished Copernicus from Ptolemy—apart from the crucial issue of whether the sun or the earth was to be deemed the center of the universe. It was also his perception that the differences that separated Copernican mathematical techniques from Ptolemy were minor, which led him to consider Greek science as being on the threshold of modern science. It is now clear that the term "trigonometry" used by Dijksterhuis actually masks a vast array of mathematical tools provided by Indian mathematical astronomers. If we continue to see their use as minor we understate grossly the vast abyss separating the mathematical instruments available to the Greeks and those available to the moderns. This raises the question of whether the avoidance of the equant by Copernicus may not also hide another great gulf dividing Greek and modern astronomy.

In order to answer this question let us begin by examining how Ptolemy deals with the problem of planetary motion. He actually gives two different models of the universe. In his *Planetary Hypotheses* he gives an account of the celestial spheres as physically real, and in the *Almagest* he gives a mathematically more precise description able to predict observational data but at the price of introducing hypotheses that violated the physical plausibility of the theory (Saliba 1994, p. 6). One hypothesis in particular was to generate controversy—his introduction of the device of the equant, which assumed the physically implausible notion that a celestial sphere can rotate uniformly about an axis that did not pass through its center. However, Ptolemy argued that since the goal of mathematical astronomy was "to save the phenomena"—that is, make accurate predictions—the physical implausibility of a theory could not be an objection against it.

However, in introducing the equant Ptolemy saw himself as remaining faithful to Plato's dictum that the motions of heavenly bodies must be explained only in terms of circular motion. But the early idea that a nested sequence of homocentric spheres centered on the earth could explain the motions, found in Aristotle's model of the universe, had already by his time turned out to be inadequate for developing a mathematically precise theory of the universe. A long tradition of better models had followed, each correcting the limitations of earlier versions. When Ptolemy came to finally treat the problem he used three different devices to construct his theory. All of them involved uniform circular motion around a fixed center in order to conform to Plato's dogma concerning heavenly motions. First, he used epicycles, which are small circles rotating uniformly about a point that is also on another circle (the deferent) rotating uniformly around its center. This epicycle-deferent system would, in general, be centered on the earth. However, when this failed to account for all observed data, even after epicycles were added onto epicycles, Ptolemy used the device of the eccentric. In an eccentric the deferent rotates around its center, but this center of rotation is displaced away from the center of the earth. Finally, Ptolemy introduced his most controversial device—the equant—to get better computational fit for his theory than any combination of epicycles and eccentrics allowed him. The equant is a point around which a deferent rotates uniformly, but it is not at the center of the deferent circle but displaced away from it. If the deferent is seen as carried on a concentric sphere, then the sphere would rotate about an axis not through its center but displaced away from it. Ptolemy's model of the universe turned out to be computationally so much better than its competitors that it soon came to dominate Hellenistic astronomy. Ptolemy's successors added epicycles, eccentrics, and equants to better align the theory with observation, but did not seek any break with his fundamental principles.

The mathematical approach of "saving the phenomena" adopted by Ptolemy did not bother the Greeks because neither Platonists nor Aristotelians believed that mathematical astronomical theory could do more. For Platonists, who saw the universe as the product of the creative activity of a Demiurge compelled to shape it out of prior recalcitrant material, there was no possibility of embodying the perfect forms of mathematics in gross matter. Hence, the Greeks under Platonic inspiration had no compulsion to suppose that perfect mathematical forms could be embodied in physical bodies (even heavenly ones)—on the contrary they would have been led to suppose otherwise. For Aristotelians who saw the creator as an "Unmoved Mover," content to contemplate himself rather than his imperfect creation, there was no compulsion to see the universe as obedient to mathematical law either. Thus both the dominant Greek philosophical schools did not lead astronomers to find objectionable the Ptolemaic view that the goal of mathematical astronomy was saving the phenomena.

However, Arabic astronomers were deeply disturbed by the notion that the goal of mathematical astronomy was merely to save the phenomena. For the Arab-Muslims the phenomena of nature were revelations of God and the created universe was a perfect unity because God, being both perfect and omnipotent, would and could embody the ideal forms of mathematics in the created world. This view even has scriptural support in the Quran, which refers to the phenomena of nature as *ayat*—that is, signs or portents of God. Indeed, according to Nasr, the same term is used for the verses of the Quran and the signs that appear within the soul of humans (Nasr 1993, p. 462). He quotes the widely known verse in the Quran:

> We shall show them Our portents (*ayat*) on the horizons and within themselves until it will be manifest unto them that it is the Truth. (XLI, 53)

Nasr argues that many Muslim thinkers also considered the cosmos to be the "Quran of Creation" or "the Cosmic Quran." They make a distinction between the Quranic revelation written in words—the recorded Quran of their scriptures—and the primordial revelation present in the cosmos whose message is reflected in every leaf and mountain, and in the sun, moon, and stars (Nasr 1993, p. 462). This view is largely inspired by repeated appeals in the Quran that the presence of order in the cosmos is evidence for the existence of God.

There is also support for the cosmos as a second revelation in the Hadith—the holiest text for the Muslims after the Quran. It is said in the *Hadith Qudsi* that "God desires to be known, so He creates the universe." Bakar interprets this verse to mean that "God's Creation is also His revelation, for otherwise it would not be possible for Him to be known through His Creation" (Bakar 1999, p. 26).

Given this spiritual orientation to the cosmos, sanctioned both by the Quran and the sayings of the Prophet in the *Hadith*, there was hardly any astronomer or natural philosopher among the Arab-Muslims who was prepared to see Ptolemy's theory as merely concerned with "saving the phenomena." To do so would be tantamount to assuming that the phenomena in the heavens were physically real in one way and mathematically appeared in a different way—that is, the cosmic revelation would have to be deemed deceptive. This led Arabic scientists and philosophers to criticize the Ptolemaic theory from the point of view of their spiritual orientation—namely that it did not conform to the requirements of physical realism. They set out to amend the theory—which they assumed to be correct in its general features—so that it would better fit their notion of a sacred cosmos revealing the power of God. Hence they demanded that mathematical astronomy should not only be physically true, but also conform to mathematical order to a degree of precision that would reflect the omnipotence of God and the perfection of his Creation. To suppose otherwise would be to assume the sacrilegious view that the revelations of the Cosmic Quran deceived—or could deceive.

As a result many Arabic scientists were led to criticize the use of the equant in Ptolemaic astronomy from a physical point of view.[8] One of the earliest important critics was Alhazen. His *Résumé of Astronomy* in Arabic has not survived, but Latin and Hebrew translations of it are available (Nasr 1968, p. 176). In this work Alhazen levels a scathing critique of the abstract mathematical approach of Ptolemy and describes how he intends to revise the Ptolemaic model at the same time:

> The movements of circles, and the fictitious point which Ptolemy has considered in a wholly abstract manner, we transfer to plane or spherical surfaces, which will be animated by the same movement. This is in fact a more exact representation; at the same time, it is more comprehensible to the intelligence … we have examined the diverse movements produced within the heavens in such a way as to make each of these movements correspond · to the simple, continuous and unending movement of a spherical body. All these bodies, assigned thus to each of these movements, can be put into action simultaneously, without this action being contrary to their given position and without their encountering anything against which they could strike or which they could compress or shatter in any way.[9]

This physical model was to exercise a great influence on European astronomers until the advent of Kepler. His attacks on the Ptolemaic abstract approach to planetary motion were continued by Andalusian scholars—including Ibn Tufail and Averroes. The final outcome of this anti-Ptolemaic movement in Andalusia was a book by al-Bitruji, *The Principles of Astronomy*, which returned to the Aristotelian model of nested concentric spheres with the earth as center to account for planetary motions without the use of epicycles, eccentrics, or equants. Each of the spheres transfers part of the motion it receives from the sphere above to the one below, with the ultimate source of motion being the outermost sphere or *primum mobile* located above the sphere of the fixed stars. With only constant and concentric motion available, al-Bitruji attempts to account for the irregular motions of the planets by means of spiral movements involving rotations of poles around poles of the concentric spheres. However, from a predictive point of view the program was a failure—more an inspiration to seek for better non-Ptolemaic models (Sabra 1984, pp.134–136).

The resolution to the problem of the equant lay in the discovery of two new mathematical theorems that have since come to be labeled the Tusi couple and the Urdi lemma. The Tusi couple was discovered by the Arabic astronomer Nasir al-Din al-Tusi. It may be formulated as follows. Consider two spheres, one inside the other and half the radius of the larger sphere, and in contact at one point with the larger sphere. If the large sphere is made to rotate about an axis, and the small sphere rotates about a parallel axis at twice the speed and in the opposite direction, then the theorem states that their original point of contact would trace a path that would oscillate back and forth along the diameter of the larger sphere. The importance of the Tusi couple in astronomical model building is that it allows a system of rotating spheres to generate linear motion, in the way turning wheels move the piston of a steam engine back and forth.

The Urdi lemma was the discovery of Muayyad al-Din al-Urdi, and is a development of the Apollonius Theorem. It states that if we have a straight line, and draw two lines of equal length from it, making equal angles either externally or internally, and connect their tops to each other, the resultant straight line will be parallel to the original line. The significance of the Urdi lemma to astronomical model building is that it allows us to transform eccentric models into epicyclic ones (Saliba 1994, p. 269).

According to Saliba both theorems played an important role in Arabic attempts to develop non-Ptolemaic astronomical theories. First, the Urdi lemma made it possible to transform any eccentric motion involving rotation about a center away from the center of a deferent into epicyclic motion, and back, so that it became possible to transfer segments of any model from the central points to the periphery and back again. This allowed astronomers to retain the effect of the equant without using the equant, and also produce uniform motions that conformed to natural physical principles. Moreover, the Tusi couple, by making it possible to produce linear motion by a combination of circular motions, allowed astronomers to enlarge and shrink the size of the epicycle radius using only combinations of uniform circular motion (Saliba 1996, p. 125).

The great flexibility for model building, rendered achievable by these theorems, allowed al-Shatir to complete his non-Ptolemaic geocentric model of the universe without the use of the equant. This became the culmination of a long critique of Ptolemaic astronomy, which began with Alhazen, continued through the so-called Andalusian revolt against Ptolemy, and developed in the Maragha School with the discovery of the Tusi couple and the Urdi lemma. Al-Shatir's planetary model had the same predictive power as Ptolemy's minus the equant. It thereby resolved the problem of the equant by eliminating the need for it. It can rightfully be described as a triumph of Arabic astronomy over its Ptolemaic predecessor.

In the 1950s E. S. Kennedy discovered that the solar, lunar, and planetary models of al-Shatir are mathematically identical to those proposed by Copernicus some 150 years later. Its significance is elaborated by Saliba:

> Namely, there seems to be a dramatic similarity between the technical results reached by Copernicus, and those reached by Maragha astronomers some two or three centuries earlier. The only distinction, of course, is the heliocentric theory of Copernicus, versus the geocentric one of the Maragha astronomers. But, mathematically speaking, and to put it in modern terminology, this reversal of the direction of the vector that connects the earth to the sun was so well known to Copernicus as to have been of no real mathematical significance ... But the real revolution in the work of the "Maragha School" astronomers lies in the philosophical dimension that was equal in importance to the mathematical and astronomical dimensions if not more so, and which was in the realization that astronomy ought to describe the behavior of physical bodies in mathematical language, and should not remain a mathematical hypothesis, which would only save the phenomena. (Saliba 1994, pp. 255–256)

Indeed so similar are the mathematical instruments and the motivations of Copernicus to those of the Maragha School that Swerdlow and Neugebauer are led to conclude in their study *Mathematical Astronomy in Copernicus' De Revolutionibus* that "In a very real sense, Copernicus can be looked upon as, if not the last, surely as the most noted follower of the Maragha School" (Swerdlow and Neugebauer 1984, pp. 295).

Thus, given the historical connections of Copernicus with Arabic astronomical traditions, we would have to reevaluate Butterfield's judgment concerning Copernicus' motives for rejecting the use of equants in astronomical theory. Butterfield writes:

> [Copernicus] had an obsession and was ridden by a grievance. He was dissatisfied with the Ptolemaic system for a reason which we must regard as

a remarkably conservative one—he held that in a curious way it caused offence by what one can almost call a species of cheating. Ptolemy had pretended to follow the principles of Aristotle by reducing the course of the planets to combinations of uniform circular motion; but in reality it was not always uniform motion about a center, it was sometimes only uniform if regarded as angular motion about a point that was not the center. Ptolemy, in fact, had introduced the policy of what was called the equant, which allowed of uniform angular motion around a point which was not the center, and a certain resentment against this type of sleight-of-hand seems to have given Copernicus a special urge to change the system. (Butterfield 1957, p. 25)

The passage correctly identifies the urge for change that Copernicus felt, but it seems to suggest that it was motivated by Ptolemy's failure to play by the rules of the game. This leads Butterfield to charge Copernicus with intellectual conservatism—perhaps he should not have taken the rules so seriously! However, the charge misses the mark since what Copernicus demanded was a radical change. He wanted an astronomical theory that would conform to physical realism—as did Alhazen, al-Bitruji, al-Tusi, al-Urdi, and al-Shatir. Such an attitude is not unnatural within the Christian conception of an omnipotent creator God—a view Catholics shared with the Muslims. It would have only been an impossible requirement in a Greek universe where the Platonic Demiurge was not omnipotent, and Aristotle's Unmoved Mover was indifferent to the universe outside He considers imperfect.

Thus, when historians of science, such as Kuhn and Dijksterhuis, suggest that the Copernican Revolution could conceivably have occurred in the second century CE, they are ignoring the extended mathematical apparatus developed by Indian mathematical astronomers, and the array of physical realist critiques and associated alternative (albeit geocentric) planetary models developed by Arabic astronomers over many centuries. Dijksterhuis' argument that, barring the use of trigonometry and the elimination of equants, Copernicus took only a small step from Ptolemy trivializes the impact of both Indian and Arabic contributions. The two words he uses—trigonometry and equant—convey the profound multicultural influences that conditioned the Copernican turn. Only by disregarding these influences can we claim that Greek science, with Ptolemy, was on the threshold of modern science.

Chapter 8

The Alhazen Optical Revolution

Another reason why we cannot assume that Greek science was on the threshold of modern science is the revolution in optical theory that intervened between Ptolemy's optical investigations, the last major important study in Greek optics, and the science of optics that shaped the thought of Renaissance artists as well as scientists like Galileo and Kepler. This new optical theory was also closely related to the discovery of the telescope, which played a seminal role in unearthing evidence crucial for the ultimate triumph of the Copernican and Newtonian theories. The optical revolution was the work of the scientist and natural philosopher Ibn al-Haytham, better known in the West as Alhazen (965–1030 CE). Alhazen's *Optical Thesaurus* had a profound influence on European optics immediately after it was translated into Latin in the twelfth or early thirteenth century. It influenced Grosseteste and Roger Bacon, as well as leading optical theorists in the Middle Ages such as Pecham and Witelo. Galileo was to appeal to the *Optical Thesaurus* to prove that the moon was not a polished mirror, as some Aristotelians maintained; and Kepler began his studies from the point where Bacon, Pecham, and Witelo had developed the Alhazen optical paradigm (Lindberg 1992, pp. 312–315).

Given Alhazen's profound influence on optical science in Europe from the thirteenth century, it is unfortunate that his role in medieval optics has been systematically ignored by historians of science concerned with the modern Scientific Revolution. This has distorted our understanding, even led to an absence of awareness, of the role played by Alhazen in paving the way to the optical revolution associated with Huyghens and Newton. It has also sustained the erroneous perception that no significant developments took place between the period when the Greeks studied optical phenomena and when optics became an important element of seventeenth-century science in Europe. Even Thomas Kuhn, the distinguished historian of science, appears to have overlooked Alhazen in his highly influential study *The Structure of Scientific Revolutions* when he writes:

> No period between remote antiquity and the end of the seventeenth century exhibited a single generally accepted view about the nature of light. Instead there were a number of competing schools and subschools, most of them espousing one variant or another of Epicurean, Aristotelian or Platonic theory. One group took light to be particles emanating from material bodies; for another it was a modification of the medium that intervened between the body and the eye; still another explained light in terms of an interaction of the medium with an emanation from the eye; and there were other combinations and modifications besides. Each of the corresponding schools derived strength from its relation to some particular metaphysic, and each emphasized, as paradigmatic observations, that particular cluster of optical phenomena that its own theory could do most to explain. (Kuhn 1970, pp. 12–13)

Kuhn proceeds to argue that optics emerged as a scientific discipline only after Newton, with his corpuscular theory of light, offered a paradigm around which a community of scientists could build a cumulative and consensually acceptable body of knowledge. Such an account completely ignores Alhazen's optical discoveries that not only managed to synthesize the attractive features of the three different ancient optical theories Kuhn refers to, but also proposed a new theory of light behavior that became the accepted paradigm of optics for most leading European scholars, including Galileo and Kepler, until the time of Newton.

In order to appreciate Alhazen's accomplishment, let us look at the state of optics as it ended with the Greeks prior to the time Arabic scholars were to take up the subject. According to Lindberg all Greek optical studies were channeled by narrowly defined criteria that guided investigators into a specific range of problems, which did not take into account the wider context of optics (Lindberg 1992, pp. 308–309). The Aristotelians were concerned only with the physical nature of light and the physical mechanisms that mediated perception between the object and the observer's eye. They did not analyze the phenomena observed mathematically, nor were they concerned with the anatomical or physiological apparatus that made perception possible. Aristotle had argued that perception occurred because the perceived object produced an alteration in the transparent medium between the object and observer—a modification transmitted instantaneously to the observer's eye to generate sensation—and Aristotelians confined themselves to problems associated with developing this theoretical perspective. Lindberg refers to this approach as adopting what he calls an "intromission" theory since the agent responsible for vision enters the eye from the outside.

An alternative to the Aristotelian theory was proposed by Greek atomists like Epicurus. Unlike Aristotle they did not consider the object to merely modify the medium between itself and the eye. Instead they argued that a thin "skin" or "simulacrum" of atoms was "peeled" off from the surface of the object and carried through the medium directly into the observer's eye. Lindberg considers this to be another kind of intromission theory since it also maintains that perception occurs because of something that enters the eye from the outside.

However, both the Aristotelian and Epicurean intromission theories were qualitative theories that explained the causes of perception but did not offer any mathematical account of the way the image in the eye was formed by the influence entering it from the outside. By contrast the mathematician Euclid, following the Platonic view that perception occurred as a result of an emanation from the eye, was able to develop a mathematical theory of perception. In his *Optics* Euclid proposed a geometrical theory of spatial perception, which assumes that radiation emanates from the eyes in the form of a cone—the visual cone—and perception occurs when the rays forming the cone get interrupted by an external opaque object. The observer identifies the object—its shape, size, and location—by the pattern and location of the intercepted rays. Euclid's views were to be further developed by Ptolemy. Lindberg refers to the theory they proposed as an "extramission" theory since perception transpires not by virtue of something entering the eye but by virtue of something emanating from it. However, although the Euclidean theory was mathematically sophisticated, it failed to take into account important nonmathematical aspects of vision—for example, it did not explain why perception could not occur in the dark if the source of illumination is radiated out of the eye of the observer.

We have seen that these three ancient Hellenic theories were considered by Kuhn to be the dominant views prior to Newton. Nevertheless, Lindberg refers to a fourth Greek theory derived from medical practitioners. This theory was inspired by studies of the anatomy of the eye and the physiology of sight and rooted in the tradition developed by the physicians Herophilus and Galen. In particular Galen (129–216 CE) made significant contributions to visual theory by giving an analysis of the structure of the eye and the various organs that formed the visual pathways facilitating vision. This theory was also nonmathematical, like the intromission theories of Aristotle and Epicurus, but its concern was more with the mechanisms in the eye that made vision possible than with the kind of signal that passed into the eye from the outside.

Lindberg argues that Alhazen's seminal contribution to optics was to produce a synthesis of these four quite different theories that Arabic thinkers had inherited from the Greeks. Alhazen's theory set out to import the mathematical sophistication of the physically implausible extramission theories of Euclid and Ptolemy into the physically plausible, but mathematically naive, intromission theories proposed by the Aristotelians and the atomists. To accomplish his task he also had to draw upon ideas from the anatomical and physiological theory of Galen. By selectively combining elements from all of these theories he was able to arrive at a comprehensive synthesis capable of explaining optical phenomena and vision with a precision and scope unprecedented in the history of the discipline (Lindberg 1992, p. 309).

Alhazen began by demolishing the extramission theory. He argued against the plausibility of the theory by drawing attention to the ability of bright objects to injure the eye—the injury must be caused by something without, since an emanation issuing from the eye cannot, by its very nature, damage the eye. Moreover, if we assume that perception is the result of such an emanation, then we are compelled to conclude that this material from the eye must fill the whole of space that can be reached by vision—extending even up the sphere of the fixed stars. This is an incredible claim lacking physical plausibility.

Having rejected all extramission theories on physical grounds he proceeds to offer a new intromission theory—a theory able, unlike the Greek intromission theories, to appropriate the visual cone of the extramission theory along with the mathematical power associated with it. Alhazen's theory was inspired by the radically new conception of radiation formulated by al-Kindi (805–873 CE), who is generally regarded as the first of the great Arabic philosophers. Al-Kindi had proposed that objects radiate light through an incoherent process—that is, each individual point, or small part of a body, radiates light in all directions and independent of other points or parts of the body (Lindberg 1992, p. 310). The process is incoherent because the radiation emitted in all directions from a single point is not coordinated with the emissions made by other neighboring points.

However, it might appear that such an incoherent process would make it impossible to perceive an object since rays from every point in the object would be received by every point in the eye, leading to either confusion or a uniform blank luminous field—not an image of the object. Alhazen solved the problem by arguing that although it was the case that every point of the eye does receive radiation from every point in the visual field, only those rays that fall perpendicularly on the eye are capable of making themselves felt. The others are not sensed because they are refracted and weakened, so that their role in vision is only marginal. He assigns a primary role to the crystalline humor or lens of the eye in effecting this process. According to him the crystalline lens of the eye only pays attention to the perpendicular rays so that each point of the object gives rise to a single ray sensed by the crystalline lens. Moreover, the perpendicular rays subtend a visual cone with the center of the eye as its vertex; and the visual field as its base.

Alhazen's approach is familiar to students today in the ray diagrams they are taught when they learn about the optics of vision. By taking the visual field as the base and the center of the lens as vertex, Alhazen managed to get the visual cone, which made it possible for the extramission theory to exploit the mathematical features of visual perception (Lindberg 1992, p. 311). However, in contrast to the extramissionists who saw the cone as involving emissions reaching out from the eye into the world, Alhazen conceived of the cone as formed by rays from the object entering the eye and sensed by the crystalline lens.

By importing the visual cone of the extramissionists into an intromission theory of perception, Alhazen manages to combine the mathematical virtues of the extramission theory with the physical plausibility of intromission theories. Moreover, by making the anatomy of the eye central to the creation of the cone, he has also incorporated the Galenic tradition of physiological optics. This was a powerful synthesis of mathematical, physical, and physiological models of perception, and soon after his theory reached Europe through translations, it came to dominate thinking about light and perception.

Hence, far from what Kuhn presumes, optics did not achieve paradigmatic status with Newton but with Alhazen. Newton may have changed the optical paradigm by introducing the corpuscular theory of light, but the paths along which these corpuscles traveled were themselves the rays that radiated from the object, as envisaged by Alhazen. Moreover, the corpuscular theory continued to retain important features introduced by Alhazen—the intromission theory, the central role of the crystalline lens, the visual cone, the analysis of perception in terms of rays are all rooted in the innovations introduced by him.

Alhazen himself was to apply his theory to a wide range of optical phenomena—the behavior of light reflected by spherical and parabolic mirrors, refraction through glass cylinders and spheres, the magnifying effect of plano-convex lenses, and the formation of shadows and the images formed in the *camera obscura* (Nasr 1968, p. 129). Even more significant for the future development of science was his deployment of the experimental method in his studies. It is well documented that he did not merely collect data for his studies passively, as and when they presented themselves in nature, but used a lathe in his laboratory to make curved lenses and mirrors on which he conducted his investigations.

In his combination of the mathematical-theoretical approach with experimental techniques, he could even be seen as anticipating important features of modern science. Alhazen's experimental approach certainly was an advance over the Greek propensity to theorize without experimentation and was to influence Roger Bacon—indeed the latter's advocacy of the experimental method was inspired by his studies of Alhazen's works. These limits of the Greek approach were contrasted with the limits of the Chinese approach by the Damascene scholar al-Jahiz around 830:

> The curious thing is that the Greeks are interested in theory but do not bother about practice, whereas the Chinese are very interested in practice and do not bother much about the theory. (Quoted in Needham 1970, p. 39)

It must not be supposed that the Alhazen achievement was an isolated phenomenon—it developed out of, and was itself a part of, wide-ranging optical studies conducted by Arabic scholars. We have seen that Alhazen was himself influenced by the philosopher al-Kindi—as the first of the Arabic philosophers, his influence cannot be underestimated. Arabic optics has to include the discoveries of al-Nayrizi's studies of atmospheric optical phenomena, the discussion on the speed of light and whether it was finite by Ibn Sina and al-Biruni, and the investigations into the anatomy and physiology of the eye by physicians such as Hunayn Ibn Ishaq and al-Razi.

However, the impact of Alhazen's optical theory cannot be confined to the domain of physics, or even science. Its theoretical discovery played a key role in the shift of sensibility in Europe associated with the Renaissance in two directions. First, it entrenched even more deeply the mathematical realism that led Arabic scientists, as we saw earlier, to object to the instrumentalist and antirealist features of the Hellenic—and more specifically, Ptolemaic—tradition. Second, it moved European consciousness toward greater trust and valuation of a sensory-observational approach to nature that is now seen as characteristic of both Renaissance art and modern science. Let us examine these changes more carefully since they can be considered to be two of the most significant transformations of consciousness associated with modernity.

Consider the swing toward mathematical realism. This is generally associated with the notion that the regularities in nature are obedient to principles that can be expressed as mathematical regularities. By contrast the Greeks held quite a different view. They did not entertain the idea that perfect mathematical forms can be embodied in the world, because even in the realm of geometry it was impossible to find perfect triangles, circles, or straight lines in concrete physical objects. Hence, an empiricist like Aristotle was inclined to adopt a nonmathematical approach to study natural phenomena in their concrete contexts. Others like Plato, who were mathematically inclined, tended to move away from the world in order to study the perfect intellectual forms. Under their influence Hellenic astronomers were satisfied if their theories could, as far as reasonably possible, "save the phenomena." To demand more—that is, to demand that the world reflect perfect mathematical order—was deemed to be asking the impossible. Even the great Archimedes found it necessary to appeal to idealized weightless pulleys, perfectly smooth inclined planes, frictionless fulcrums for levers, and so on. Mathematics could be applied only to ideal objects, but in the case of real objects mathematical applications could at best only be approximate.

Compared with the mathematical idealizations used in Greek science, light rays in Alhazen's optics were real rays that could be deemed to travel in perfect straight lines. There was no separation here between the physical object studied and the mathematical laws it obeyed through the mediation of an ideal object. His optics became the first example of the application of mathematics to physical phenomena in which the phenomena themselves could be treated as embodying perfect mathematical relations. It served as the archetype of a mathematical realist theory of events in the world. This is what inspired Roger Bacon and Robert Grosseteste to treat it as the paradigm of what a scientific theory should be, and motivated, to a large extent, their attempt to develop its wider methodological implications for knowledge in general by introducing an early concept of the experimental method.[1] Their writings played an important role in guiding others within Europe along a path that ultimately led to modern science.

Second, Alhazen's optical theory also led to a profound shift in the perceptual sensibilities of Europeans. It did this by transforming the notion of what it is to make unbiased and accurate observations. Whereas Greek art was concerned with the representation of ideal types or forms, and Christian art in Europe was directed toward creating symbolic representations of spiritual ideas and events, the new optical science gave birth to a theory of perspectivism in which the artist could represent objects as they appeared to the eye of the observer. To represent the world as it is or as it appeared to the eye of the observer—without the interposition of idealizations or spiritual projections—became a real possibility and excited Renaissance artists. At least this is what Renaissance artists saw themselves doing once they accepted the account of perception given by the theory. The world now could be considered to have become transparent to the human visual gaze.

The transformation of Renaissance consciousness in Europe from the intellectual gaze of ancient Hellenism, and the visionary gaze of medieval religion, to the visual gaze of the moderns has been explained in many ways. However, it could be argued that one important factor, if not the main one, in this change was the new theory of vision embodied in the Alhazen optical revolution.[2] Unlike Platonic theories where the eye radiates outward to capture the object, or Aristotelian theories where the eye's perception of the object is mediated through the modifications of the medium between, the Alhazen theory suggested that every point in the object has a one-to-one correspondence with a point in the image sensed in the crystalline lens. The image becomes an exact representation of the world outside—and art, by becoming mathematical, could capture it on canvas. This led many artists to study the ways in which rays from objects reached the eye of the observer—the mathematical science of perspectivism became the foundation for the art of the Renaissance. The goal of art was representation of the visual field. This representational paradigm came to dominate European aesthetics until the end of the nineteenth century.

The notion of exact and accurate observation of nature that inspired the artists—along with the confidence given by the Alhazen theory that such observations did not deceive because they were underpinned by the correct understanding of how the human eye perceived the world—also had an impact on science. It generated the confidence that accurate, unbiased observation of nature was both desirable and possible in science as in art. The link between the Renaissance artist and scientist has been noted by Butterfield:

> Finally, the Renaissance brought a greater insistence on observation and a refinement of observational skill. It is perhaps not an accident that the first branch of science transformed by improved observation was that of anatomy, the science of the painter, the one restored by Vesalius, in whom the mind of the artist and the mind of the scientist seem almost to have been fused into one. (Butterfield 1957, pp. 52–53)

However, the impact of Alhazen's theory was to also reach forward into modern Enlightenment thought. In his book *Downcast Eyes: The Denigration of Vision in Twentieth-Century French Thought,* the historian Martin Jay describes attempts by postmodern thinkers to deconstruct the epistemology of Enlightenment thought inspired by appeal to visual metaphors. This movement includes influential figures such as Jean-Paul Sartre, Michel Foucault, and Jacques Derrida. Jay argues that it was Descartes, the father of modern philosophy, who first enthroned vision as the master sense in modern thought. He maintains that Descartes laid the basis for the visualist paradigm of knowledge in his treatise *La Dioptrique.* Jay traces the roots of the Cartesian epistemological orientation, which he labels "ocularcentrism," to perpectivalist painting:

> [Descartes] tacitly adopted the position of a perspectivalist painter using a camera obscura to reproduce the observed world. "Cartesian perspectivalism," in fact, may nicely serve as a shorthand way to characterize the dominant scoptic regime of the modern era. (Jay 1993, pp. 69–70)

As a result of the Cartesian influence, even conceptual thinking came to be dominated by the visual paradigm of knowledge—one sees thoughts in the mind in the same way one sees objects in the visual field. This has been criticized by Rorty, whose book *Philosophy and the Mirror of Nature repudiates* of the notion that our conceptions of the world somehow mirror reality. Rorty writes:

> In the Cartesian model, the intellect inspects entities modeled on retinal images... In Descartes conception—the one that became the basis for modern epistemology—it is representations which are in the "mind." (Rorty 1980, p. 45)

These views parallel recent developments in the philosophy of science associated with Hanson, Toulmin, Kuhn, and Feyerabend, which also attempt to break away from the ocularcentric or visualist paradigm. Central to the views of these thinkers is that all observation is theory laden—a view inspired by discoveries associated with gestalt psychology, historical studies in science, anthropological studies, and philosophical analysis of perceptual experience. They conclude, as Hanson puts it, that "there is more to seeing than meets the eyeball." (Hanson 1961, p. 7) They maintain that since perception is dependent on conception, it puts into question traditional empiricist epistemologies that naively assume theories can be directly inducted from the phenomena, or uncritically confirmed or disconfirmed by observation.[3]

Nevertheless, in spite of the important role played by Alhazen's optical theory in shaping the ocularcentric turn within modern epistemology criticized by the postmodern thinkers above, there has hardly been any historical recognition of his influence. The main problem appears to be the dominant Eurocentric orientation of historical studies in science, which makes it difficult to accommodate and acknowledge the dialogical influences that have conditioned both modern science and its philosophy. If the arguments presented above are accepted, and the revolution in optical theory by Alhazen did have a profound impact in shifting aesthetic, scientific, and epistemological sensibilities in the modern era, then we have one more reason to suspect that Greek science could not have been on the threshold of modern science. Moreover, the route to modern science cannot be understood by ignoring the important role played by Arabic science and philosophy in paving the way for it.

Nevertheless, our account leaves one important question unanswered: If the Alhazen theory had such an important impact on Europe, both by way of transforming aesthetic sensibilities, and creating the ocularcentric perceptual and conceptual reorientation to nature, why did it not have the same effect within the Arabic world? After all, Arab-Muslim artists should have been even more receptive to the mathematical orientation to nature that this theory inspired. Indeed the sacred art of Islam was an abstract mathematical art using regular geometric figures interlaced with one another, and combining flexibility of line with emphasis on archetypes. According to historian of science Syed Hossein Nasr, the use of mathematics in art appealed strongly to Muslims, because its abstract nature furnished a bridge between the multiplicity of the world and the Divine Unity they worshipped. The answer has to be that in Islam all representation of the natural world was forbidden—or held to be strongly suspect. Hence the latent possibilities of Alhazen's optical theory that were to have profound consequences in Europe were never realized in the Arabic world.

Chapter 9

The Modern Atomic Revolution

The atomic theory, as it emerged in the seventeenth century, involved a radically new philosophy of nature that broke the hold of the Aristotelian tradition of science by proposing that all phenomena in the physical universe can be explained in terms of the motion of tiny indivisible particles moving in a void space. It violated the fundamental assumptions of Aristotle's physics in two important respects—it rejected the Aristotelian principle of the indefinite divisibility of matter and the view that space is a *plenum* in which the existence of a void was impossible because nature abhorred a vacuum. An atomic conception of nature was fairly widespread among the founders of modern science—Bacon, Gassendi, Boyle, and Newton all subscribed to some version of corpuscularism. The general understanding of the change to the corpuscular philosophy is that it took its inspiration from the atomists of the ancient Hellenic and Hellenistic world—Leucippus, Democritus, Epicurus, and Lucretius. It is also assumed that atomic views of nature were hardly discussed in Europe between the time of Lucretius and its reemergence as an alternative framework to Aristotelianism for natural philosophy in the seventeenth century.[1] Of course, even in the ancient tradition of science and philosophy, the atomic view was rejected by the much more influential Platonic and Aristotelian schools.

However, this conception of the history of atomic ideas is questionable. While it is true that atomism was never favored or adopted by any important European thinker after the ancient atomists, the atomic model as an alternative conception to Aristotelian metaphysics must have been well known to medieval European scholastics. They could not have avoided it given that an atomic conception of nature constituted the dominant school—if not the only school—of Islamic theology during the medieval era in Europe, and continues to be taken seriously even today by Sunni theologians. Going by the name of *kalam* this Arabic tradition adopted a different conception of atoms from the Hellenic thinkers. *Kalam* thinkers argue that atoms have only a momentary existence—they appear instantaneously and then disappear only to be replaced by other atoms of a similar nature.[2] Labeled as atomic occasionalism, this view differs from the Hellenic view of atoms as indestructible and eternal.

One important route through which European thinkers could have become acquainted with *kalam* is the writings of the Jewish scholar Moses Maimonides. His magnum opus, *The Guide to the Perplexed*, was originally inspired by his attempt to demolish *kalam* atomism and defend Aristotelian natural philosophy. In his study, which was widely read by leading European thinkers and theologians, including Thomas Aquinas after it was translated from Arabic into Latin, Maimonides discusses and criticizes *kalam* views in great detail. He identifies twelve propositions shared by the *mutakallemim*, as the *kalam* proponents were known. The first two of these are directly relevant to our concerns here—namely that all things are composed of atoms and that there is a vacuum. Both violate fundamental principles of Aristotelian natural philosophy.[3]

Because Maimonides was writing at a time when the debates between Islamic atomists and Aristotelians was at its height, his views would surely have been of intense concern to European thinkers who wanted to defend Aristotelian philosophy as compatible with scriptural revelations—a task often considered to have culminated with Thomas Aquinas. According to Maimonides the first and central claim of the atomists is as follows:

> The universe, that is, everything contained in it is composed of very small parts [atoms] which are indivisible on account of their smallness; such an atom has no magnitude; but when several atoms combine, the sum has a magnitude and thus forms a body ... These atoms, they believe, are not, as was supposed by Epicurus and other Atomists numerically constant; but are created anew whenever it pleases the Creator; their annihilation is therefore not impossible. (Maimonides 1956, pp. 120–121)

The second most significant claim made by the school of *kalam* is the doctrine affirming the void, which Maimonides presents as follows:

> The original *mutakallemim* also believe that there is a vacuum; i.e. one space, or several spaces which contain nothing, which are not occupied by anything whatsoever, and which are devoid of all substance. This proposition is to them an indispensable sequel to the first. For if the universe were full of such atoms, how could any of them move? And yet the composition, as well as the decomposition of things can only be effected by the motion of atoms! Thus the *mutakallemim* are compelled to assume a vacuum in order that the

atoms may combine, separate and move in that vacuum which does not contain any thing or any atom. (Maimonides 1956, p. 121)

Hence, given that these central notions of *kalam* were available to European scholars through the works of Maimonides, one could hardly consider them not to have been of interest to European thinkers since they constituted central elements of the dominant school of Islamic thought opposed to Aristotelian philosophy. It is likely that atomism came to be taken seriously in modern Europe because of the significance attached to it by the proponents of *kalam* as a counter to the Aristotelian tradition, than because of the influence of the long marginalized atomic tradition of the Greeks. But the translations of classical atomic views could have made it easier for European thinkers to adopt atomism as an alternative to the Aristotelian philosophy without appearing to associate with the tradition linked to theological Islam.

Another reason for suspecting that the *kalam* atomic views could not have been unknown to medieval European scholars is the profound influence exercised by the Arabic philosopher Averroes. His defense of Aristotle against *kalam*, and the critiques of the Aristotelian tradition by al-Ghazali, led Dante to single him out with the tribute *"che il gran commento feo"* — "he who wrote the grand commentary." Averroes' most renowned work *The Incoherence of the Incoherence* was a response to al-Ghazali's critique of the Peripatetics titled *The Incoherence of the Philosophers*. Averroes' attack on the Islamic theological schools, his defense of rationalism, and his insistent call to separate religious truth from philosophical truth met with enthusiastic support in all the major new universities of Europe, largely because it was perceived as an attack on clerics in general. The Catholic Church was so alarmed by the spread of his teachings that his views came to be explicitly condemned by ecclesiastical authorities in Paris in the year 1277. The Church had earlier sent Thomas Aquinas to Paris to check the anticlerical movement, only to later condemn him for suspected Averroist inclinations, thereby placing a shadow that hung over his name for many years after his death.

Moreover, Averroism continued to spread over the next three centuries and remained an influential force in Padua even at the time Copernicus, Galileo, and Harvey studied there.[4] Given the widespread and deep authority of Averroism within European intellectual circles—even though its call to separate religion and science is now interpreted in Europe as a defense of philosophy from Catholic, rather than Islamic, clerical control—it is hardly likely that European thinkers reading Averroes would not have come to know of *kalam* atomic views of nature, which are precisely the central doctrines Averroes is intent on refuting. In fact, they could not avoid coming to know about *kalam* doctrines simply because a great portion of

Averroes' critique involves an initial description and faithful presentation of al-Ghazali's position.

This raises an intriguing question. If Greek atomic doctrines became available in Europe through translations in the medieval period at more or less the same time as the *kalam* atomic ideas, should we consider the emphasis on the impact of Greek thought to be exaggerated, and that of *kalam* ignored, by histories of science that treat the former only as the source of modern atomic views? Could *kalam* atomism have exercised a more profound influence than hitherto suspected? After all, unlike the ancient atomic views that held only a marginal position in the dominant tradition of Hellenic thought shaped by Aristotle and Plato, *kalam* constituted the dominant alternative to Aristotelian (and in general Hellenic) philosophy within Arabic culture. Even those who might want to deny that there was any influence from *kalam* would have to explain why this was so, since two of the most influential figures on medieval European thinkers, Maimonides and Averroes, wrote philosophical works whose principal objective was to defend Aristotelian rational philosophy against attacks by *kalam* atomists.

We must remember that Averroes' defense of Aristotelianism against al-Ghazali, *The Incoherence of the Incoherence,* was translated into Latin and published in Venice in 1497, and republished in 1527. Is it reasonable to suppose that later thinkers such as Gassendi (1592–1655), so influential in promoting atomic ideas against the Aristotelian worldview in the next century, were not aware that *kalam* atomism was also intent on subverting Aristotelian philosophy? The connection would have struck Gassendi even more because he too saw atomism as more in harmony with his religious beliefs and Christian teaching. He argued that atoms were created by God and of a finite number—a view later adopted also by Newton. In this respect atomism came to be linked with religious belief as it was in the Islamic world through *kalam*.

One reason for thinking that the influence of *kalam* cannot be important is the deafening silence about it within Europe at that time. But there could be an important reason for this—the close association of *kalam* with Islamic theology. Given the tensions between Christian Europe and the Islamic world, it may not have been politically correct to link the European atomic critiques of Aristotelian philosophy with similar critiques of Aristotelianism made in Islamic theological schools, especially when Aristotelian philosophy had become closely linked to the Church from the time of Aquinas. After all, the Muslims had been expelled from Spain only recently; the memories of the fall of Constantinople to the Turks in 1453 was still fresh; and their continuing military threat to Europe could not be overlooked. The silence is probably designed to insulate atomism against any links with the Islamic clerical tradition, and to absorb and domesticate it into Christian Europe as an alternative to Aristotelian philosophy.

However, it is also important to note the differences between modern and *kalam* atomic views. The atomic doctrines that emerged in the seventeenth century assumed that atoms persisted over time and had real magnitude. This notion of atoms was different from the momentary point atoms of *kalam* and closer to the views of the ancient atomists. The silence may reflect the fact that the modern theory of atoms also had significant differences with *kalam* atomic views. Hence it makes more sense to connect modern atomism with classical atomism. However, this does not mean that the pioneers of modern atomism were unaware of the importance of atomic ideas in the Arabic world as a framework of resistance to Aristotelian thought, or that they were uninfluenced by this knowledge.

Finally the silence concerning *kalam* could also have been related to the perception that the proponents of *kalam* denied the existence of secondary causes in the world, without which scientific knowledge could not be developed. According to *kalam* thinkers all events in the world are directly caused by God—not by any intermediate cause. The appearance and disappearance of any point atom is not conditioned by the appearance and disappearance of other atoms but directly attributable to the Will of God. Hence *kalam* thinkers denied that one event in the world could be the cause of another. By contrast most early modern thinkers held that although God is the primary cause of any event, the same event could have a secondary cause in another event. For example, it could be held that the secondary cause of cotton becoming ash is fire, but that fire and the laws of nature that cause fire to burn cotton both exist because God, as their primary cause, made them what they are. The *kalam* thinkers denied this: for them all processes in nature are the products of instantaneous point atoms, and the appearances and disappearances of any one of these does not influence the appearance and disappearance of another. Hence, there are no secondary causes. Instead *kalam* explains all natural changes in terms of "What God Wills" or "Because God So Wills." Thereby the proponents of *kalam* consider themselves to be magnifying God's power and action in the world, since there are no secondary causes mediating the Will of God. Everything can be explained by God's Will; nothing is the cause of any other thing—God is the only cause, and only explanation, of all processes observed in nature. Given this position of *kalam* the proponents of atomism at the dawn of modern science might have considered that any association of their ideas with it would subvert their endeavors to construct a causal system of science where phenomena are explained through secondary causes—even if it is also acknowledged that these causes have their primary cause in God. Even so, we may suspect that they came to consider the possibility of atomism as an alternative metaphysics to Aristotelianism because of the impact and influence of *kalam*.

However, from the point of view of creating a new science after the collapse of Aristotelian philosophy, it was wise to reject the *kalam* denial of secondary causes. Indeed many thinkers today concerned with explaining the decline of Arabic science in the medieval era link it to the rise of the theological *kalam* as the all-pervasive, definitive interpretation of the Islamic scriptural heritage. Thus argues Hoodbhoy:

> The decline of science in Islamic culture was contemporaneous with the ascendancy of an ossified religiosity, making it harder and harder for secular pursuits to exist. This does not pinpoint the orthodox reaction against science as the single cause. In particular, it does not exclude economic and political factors. But certainly, as the chorus of intolerance and blind fanaticism reached its crescendo, the secular sciences retreated further and further. Finally, when the Golden Age of the Islamic intellect ended in the 14th century, the towering edifice of Islamic science had been reduced to rubble. (Hoodbhoy 1991, pp. 95–96)

Hoodbhoy quotes the admonition to young scholars by al-Ghazali to support his conclusion:

> O youth, how many nights have you remained awake repeating science and poring over books and have denied yourself sleep. I do not know what the purpose of it was. If it was attaining worldly ends and securing its vanities and acquiring its dignities and surpassing your contemporaries and such like, woe to you and again woe.[5]

However, al-Ghazali's role is more complex than this simplistic account suggests. On the one hand it is true that al-Ghazali destroyed science in the Arabic world—at least the conception of science as held by the Aristotelians. Since he offered no alternative he can be said to have had a negative impact on the growth of science in the Arabic world. On the other hand it is also possible to see al-Ghazali's epistemological critique of Aristotelian science as clearing the way for modern science to emerge on its debris—and even, as we will find, contributing to the new epistemological orientation of modern science. In short, it not only destroyed Greco-Arabic science but opened the door for modern science.

This shift was brought about by al-Ghazali's destruction of the notion of the cause-effect relationship in Aristotelian science and philosophy. Aristotelians maintained that there was a necessary connection between a cause A and its effect B and that it was the goal of science to find this linkage. However, the advocates of *kalam* found the concept of causal necessity repulsive because it seemed to violate the omnipotence of God. It suggests that God could not have made the world differently from the way it is now—that is, God could not have made a different world, in which a cause A that now gives rise to its effect B in this world ceases to do so in the other world. This led al-Ghazali to develop a critique of the Aristotelian conception of causality—a critique that was to ultimately subvert the epistemological foundation of Aristotelian science. He argued that philosophy cannot rationally establish the necessity of cause-effect relations since these do not follow from the nature of things but from God's Will acting in the world. In short, even with the same nature possessed by a cause A, God could have made a world in which B did not follow from A.

In his *Incoherence of the Philosophers* al-Ghazali develops his powerful critique of the Peripatetic conception of the cause-effect relation. He writes:

> According to us the connection between what is usually believed to be a cause and what is believed to be an effect is not a necessary connection; each of the two things has its own individuality and is not the other, and neither the affirmation nor the negation, neither the existence nor the non-existence of the one is implied in the affirmation, negation, existence, and non-existence of the other—e.g., the satisfaction of thirst does not imply drinking, nor satiety eating, nor burning contact with fire, nor light sunrise, nor decapitation death, nor recovery the drinking of medicine, nor evacuation the taking of a purgative, and so on for all the empirical connections existing in medicine, astronomy, the sciences and the crafts. For the connections in these things is based on a prior power of God to create them in a successive order, though not because this connection is necessary in itself, and cannot be disjoined— on the contrary, it is in God's power to create satiety without eating and decapitation without death and so on with respect to all connections.
>
> The philosophers however deny this possibility and claim that, that is impossible. To investigate all these innumerable connections would take too long, and so we shall choose one single example, namely the burning of cotton through contact with fire; for we regard it as possible that the contact might occur without the burning taking place, and also that cotton might be changed into ashes without any contact with fire, although the philosophers [Aristotelians] deny this.[6]

This is a devastating attack on the foundations of Aristotelian science. For by undermining the necessity of the link between a cause and its presumed effect, it also subverts the Greek notion that by knowing the nature of a thing A—its essence, so to speak—we can infer what its effect would be. Following Aristotle, the Arabic *falsafah* (philosophers) had separated all causes into four kinds—the material, the formal, the efficient, and the final. Any effect could be explained by showing how the effect *necessarily* followed from this fourfold order of causes. Indeed such a demonstration was what was meant by explanation within the Greco-Hellenic tradition of science largely shaped by Aristotle. By presenting powerful arguments adopting the tools of rational philosophy to show that God could have created another possible world in which a given cause (or set of causes) may occur, and yet lead to a different effect from the one actually observed, al-Ghazali destroyed the epistemological foundations of Greco-Arabic science. The essence of a thing cannot explain its effect—there is no necessary link between a cause and its effect.

Since establishing such necessity was the goal of science in his time, the shock of his demonstration of the incoherence of the philosophers was to reverberate throughout Arabic civilization—a shock from which it never recovered. With him a whole tradition of Hellenic–Arabic science may be said to have died. After him interest in science declined and Arabic civilization turned inward toward theological concerns. Even in those few areas where science continued to advance, for example, mathematical astronomy and optics, it was largely not inspired by Aristotelian philosophy. Instead the Hellenic–Arabic tradition was passed on to the Europeans, who did not develop it but rather used it as a stepping stone to create a new tradition of science with an entirely different notion of causality—one able to accommodate the Ghazalian destructive dialectics largely by taking for granted his conception of the contingency of the cause-effect nexus.

In order to appreciate the impact of al-Ghazali's critique it is important to understand the role that was assigned to reason by the Greeks. In the Platonic tradition, reason was not only a method for proceeding deductively from well-established principles to others by application of logic, but also a process of proceeding dialectically in order to establish with certainty the first principles of a science. Hence, by combining deductive and dialectical reason, Platonists considered it possible to not only establish scientific first principles but also demonstrate other claims by deducing them from these principles. In contrast to this rationalist approach Aristotle emphasized the importance of prior empirical knowledge in developing an understanding of phenomena in the world. This difference between the two approaches recommended by Plato and Aristotle can be traced to their different metaphysical conceptions of what the being of a thing is—for Plato, things in the world are imperfect embodiments of independently existing ideal forms that had to be apprehended by the intellect; for Aristotle, the forms did not exist in any transcendent realm but could only be apprehended within particular objects. Having no independent existence they had to be known by direct observation. It is through perceptual contact with objects that we come to know the forms and arrive at the first principles of a science. However, Aristotle, like Plato, maintained that these first principles should be self-evident—hence, they cannot have any possibility of being false.

Al-Ghazali's argument against this approach to knowledge can be formulated as follows. Since every argument needs to start with premises that have to be taken for granted, the only way of beginning a science from which we can deduce cause-effect relations is by assuming some first principles. These can be established, following Plato, by appeal to reason, or following Aristotle, by appeal to experience. Suppose that either reason or experience can establish the first principles as self-evident. Such principles can then be used to infer the cause-effect linkages for phenomena in their domain of applicability. But the cause-effect relations so established would be necessary relations if they follow deductively from first principles that are self-evident. But, argues al-Ghazali, we have established that they are not necessary, but contingent, relations dependent on the Will of God. Hence the first principles cannot be self-evident. Therefore, neither reason nor experience can establish *self-evident* first principles. Moreover, if the first principles are not self-evident, then they can be doubted. Thus doubt infects knowledge precisely at the point where philosophers have assumed that doubt is impossible—in the foundational first principles of a science.

Al-Ghazali himself defines what indubitable knowledge has to be. He describes it as "that in which the thing known is made so manifest that no doubt clings to it, nor is it accompanied by the possibility of error and deception, nor can the mind even suppose such a possibility."[7] Clearly indubitable knowledge is self-evident because the thing known manifests itself as beyond doubt. However, armed with this definition, he argues that we can find certain knowledge neither in the deliverance of the senses nor in that of reason (Bakar 1999, p. 48).

Having introduced doubt into all the sciences al-Ghazali introduces a new idea to replace the Aristotelian and Greek conception of cause-and-effect connections—an event A is regularly connected to an event B not because of any necessity in the link between A and B but because God wills it so. The regularities of the connections we perceive in the world are, according to al-Ghazali, "the habits of God." Another way of expressing the links perceived between what we think of as a cause and what we think of as its effect is to see them as following the laws laid down by God—they are the products of God's Will. It is God's habits that make for the regularities in the causal linkages we perceive in the world. The regularities observed in nature express the Will of God and not necessary connections between a cause A and its presumed effect B that follows from the nature of A.

It is this Ghazalian conception of cause and effect that we find later in Newton's view that God created atoms and laid down the laws that mediate their interactions with each other:

> It seems probably to me, that God in the beginning formed matter in solid, massy, hard, impenetrable, moveable particles of such sizes and figures and with such other properties and in such proportion to space, as most conduced

to the End for which he formed them ... And therefore, that Nature may be lasting, the changes of corporeal things are to be placed only in the various separations and new associations and motions of these permanent particles ... it seems to me farther, that these particles have not only a *vis inertiae* [inertial force], accompanied with such passive laws of motion as naturally result from that force, but also that they are moved by certain active principles, such as is that of gravity, and that which causes [chemical] fermentation and the cohesion of bodies.[8]

In a way the Newtonian view is a vindication of *kalam* against Averroism. It breaks away from the Aristotelian metaphysics of the continuum and necessary cause-effect relations (even if these have to be first discerned through experience). It combines atomism with the notion that causal relations are contingent on God's Will. Of course there are important differences—Newtonian atoms are "lasting," as he puts it, and obey regularities that are laid down by God in a manner that, as Newton says, "most conduced to the End for which he formed them." Hence, Newtonian atomism views God as the primary cause who laid down the regularities that we discover as secondary causes in a universe of atoms and the void.

In order to understand the shift from the *kalam* to the Newtonian conception of atoms and the laws that govern them, we have to recognize the intermediate role played by medieval scholastic philosophy. The Newtonian belief in secondary causes can be traced to Aquinas, who made a place for Aristotelian philosophy and science by rejecting al-Ghazali's repudiation of secondary causes; and made a place for the omnipotence of God, who could have created a universe obeying other regularities, by rejecting the Averroes defense of necessary self-evident first principles in science. In place of these two views Aquinas offers a more nuanced position that combines primary (God-based) causes with secondary (creaturely) causes. This intermediate position of Aquinas between Averroes and al-Ghazali is described by Fakhry. He argues that Aquinas first stripped away the "deterministic implications" of Aristotelian philosophy by affirming the primacy of God in all causation, to make it compatible with revealed theology. Second, he taught that "the world of nature is subsistent and real in its own right" so that it could develop in accord with the laws of its own being. Fakhry quotes Aquinas as saying that

[I]t is part of the design of God's providence to allow the operation of secondary causes, in order that the beauty of order may be preserved in the universe ... and (in order that God) may communicate to creatures the dignity of causality. (Fakhry 1958, pp. 19–20)

However, Aquinas continued to defend Aristotelian philosophy and science within his metaphysics of primary and secondary causes. It required him to reinterpret Aristotle as purely an empiricist—a tradition now reflected in the oft-held view of Aristotelian empiricism as the contrast to Platonic rationalism. But it is more accurate to see Aristotle as maintaining the view that although the first principles of a science have to be discovered through experience, they become justified by virtue of their self-evidence. By defending the notion of necessitarianism for the first principles of any science, Averroes remained more faithful to Aristotle's views than those who now treat him only as an empiricist. Aristotle can be described as an empiricist in the context of discovery of the first principles of a science, but he has to be seen as a rationalist in the context of their justification. Aquinas naturalized Aristotle into the religious context of Christian Europe, and its conception of a monotheistic omnipotent God, by including Aristotelian science within a wider context that treated the causes described by Aristotle as secondary causes laid down by the primary agency of God. Being authored by God, but not necessitated upon God, the principles that defined cause-effect relations in the world could only be discovered and justified by observation—not reason.

In the seventeenth century, following the collapse of Aristotelian philosophy and science, the distinction made by Aquinas between primary and secondary causes came to be the basis for reconciling the new mechanical philosophy with religious belief. One of the leading figures who guided this transition was Pierre Gassendi. He set out to create a new metaphysics of atomism more compatible with the mechanical philosophy than the Aristotelian worldview, and that did not violate divine omnipotence. Instead of treating atoms as eternal he had God create them; instead of making them transient he invoked the notion of stable atoms created by God with properties that allowed them to act as secondary causes of events in nature. This did not preclude the fact that their existence and activity depended directly on God—the omnipresent and omnipotent primary cause.[9]

However, Gassendi held that the only way we could come to know the regularities and laws God imposed on atomic behavior is through empirical investigation, since God could have freely made them obey other laws. Hence, the shift from the Aristotelian conception of necessary causes and its rejection of atomism to the Ghazalian conception of atomism with only primary causes, to the Aquinian view of primary and secondary causes without atomism, and then to the Newtonian view of primary and secondary causes with atomism has to be seen as the outcome of a long-drawn-out debate in which Greco-Arabic science collapsed and modern science was born. To ignore these dialogical exchanges that mediated this scientific and philosophical reorientation is to impose a Eurocentric structure on the dialogical history of science.

In the eighteenth century David Hume examined the problem of causality within the framework of the new science concerned with the study of phenomena obeying laws derived by induction. He argued that there is no necessary connection between a cause and its effect, and concluded that only custom and habit lead us to consider the causal nexus necessary. Hume's conclusions can be interpreted as seeing the results of al-Ghazali's epistemological revolution from the other side—after the emergence of a science that has absorbed al-Ghazali's critique in its approach to the study of nature. Like al-Ghazali the new science saw the cause-effect links established by induction as contingent—they can be empirically discovered, but are not self-evident because the universe could have obeyed other regularities. However, Hume does not see these contingent regularities as supported by the habits of God—he translocates the support from heaven to earth by arguing that our belief in these regularities is supported by human custom and habit.

Indeed the arguments presented by Hume when he develops his critique of causality not only parallel those of al-Ghazali but even cite some of the same examples to illustrate his position. This has been noted by the historian of science Osman Bakar:

> Interestingly enough, in his repudiation of causality, Hume presented arguments very similar to those offered by the Asharites, but without positing the Divine Will as the nexus between two phenomena which the mind conceives as cause and effect. Moreover, some of his examples were the same as those of the Asharites. This led certain scholars to assume that Hume must have been acquainted with Asharite atomism through the Latin translations of Averroes' *Tahafut al-Tahafut* [*The Incoherence of the Incoherence*] and the above mentioned work of Maimonides [*The Guide to the Perplexed*]. (Bakar 1999, p. 101)

However, whether we see the world in terms of atoms that obey laws that are habits of God, or habitual ways of organizing knowledge by human beings, the conception of such regularities as involving necessary connections has died. Today *kalam* atomism and the epistemology of causal contingency it espoused to defend the omnipotence of God may have faded away in modern science, and the regularities it saw as the habits of God may have come to be seen as only human habits. However, even with this change traces of the Ghazalian impact remain. No longer can we return to the Greco-Arabic vision of a science based on self-evident first principles; our science is based on laws that could have been otherwise. Since *kalam* paved the way to modern science by destroying the epistemological foundation of Greco-Arabic science, can Hellenistic science be said to have been on the threshold of modern science? Was the Renaissance in Europe nothing more than the rebirth of Greek science and philosophy in modern Europe?

Chapter 10

Integrating Hellenic and Indian Traditions

The view of Arabic civilization as merely transmitting back to Europe the tradition of Hellenic science and philosophy also ignores the significant synthesis made by Arabic scholars of the traditions of thought they inherited from the Greeks, Persians, Indians, Syrians, and others who became a part of the Islamic empire. However, the Arabic historians themselves were careful to record their indebtedness to these predecessor traditions. It is also well known that under the Sassanian Empire founded in the third century by Shapur I, there was a long tradition of study of the ancient sciences of various civilizations by scholars from diverse cultures who gathered together in Jundishapur in Persia. The school of Jundishapur continued the tradition of learning after the conquest of the Persians by the Muslim Arabs.[1]

As a result Indian, Persian, and Greek sciences were cultivated in the Arabic lands, and one of the great accomplishments of the Arabic civilization was the creative fusion of these traditions in many areas of knowledge. In disciplines such as zoology, anthropology, and chemistry the Indian and Persian traditions were dominant; in mathematical astronomy the Indian tradition was considered the most advanced by the Arabic scholars because it had built on the Greek tradition and taken computational techniques much further; and in some disciplines like alchemy the Chinese influence was significant. In the area of philosophy—among the *falsafah* thinkers—and in the development of mathematical geometry and optics the Hellenic tradition was considered preeminent.

According to Needham one important factor that may have led European scholars to ignore the impact of Indian and Chinese cultures on Arabic thinkers may be the selective way medieval European translators drew on Arabic texts. He writes:

> But for some reason or other, when the translations were being made from Arabic into Latin, it was always the famous authors of the Mediterranean antiquity who were chosen, and not the books of Islamic scholars concerning the science of India or China .... As early as the middle of the +9th century Ali al-Tabari, son of a Persian Christian astronomer, who lived in Baghdad, wrote his great medical work *Firdaus al-Hikma (The Paradise of Wisdom)*, and it is striking that he quoted from Indian physicians such as Caraka, Susruta, and Vagbhata II, no less than from Hippocrates, Galen and Dioscorides. But after a thousand years al-Tabari's work has not yet been translated into a western language. Similarly the great al-Khwarizmi, whose work *Hisab al-Jabr wa'l Muqabalah*—on algebra—was written about +820, introduced a knowledge of the Indian numeral system. Fifty years earlier, al-Fazari had certainly been acquainted with parts at least of the Indian astronomical work *Surya Siddhanta*.
>
> Representative of this trend stands above all the great al-Biruni, who having followed Mahmud of Ghaznah in his conquest of India, returned and wrote very early in the +11th century, about +1012, his admirable work *Ta'rikh al-Hind.* This is not only a history and geography of India in the ordinary sense, but a profound examination of all the sciences of the Indians. But it was not translated into any European language until 1888! (Needham 1970, pp. 15–16)

One effect of this selective process of translation is to make European historians, when they came to write the history of modern science and the Arabic impact upon it, ignore the wider context of the philosophical and scientific ideas they inherited from the Arabic tradition. They were able to locate and take into account the Greek roots of these ideas but not other influences.

These factors, peculiar to the historical and theological context of the assimilation of Arabic works, have grossly distorted our understanding of Arabic philosophy and science. For one thing, only the *falsafah* thinkers are treated as having developed a philosophy of nature, but the *kalam* thinkers, who can be said to be more acquainted with the Indian philosophical tradition, are marginalized. Second, the role of the Arabic scholars as mediators of Greek and Indian medical and mathematical knowledge to Europe is widely recognized, but the role they played in integrating Greek and Indian medical traditions, as well as in uniting Greek geometry with Indian algebra, is often underestimated. This makes it possible to promote them as carriers and transmitters of the Hellenic heritage to Europe (with some Indian ideas), but completely overlooks their achievement in bringing about an accommodation between the different traditions they had inherited.

This is extremely convenient for the construction of Eurocentric histories of Arabic science as merely having the role of preserving and transmitting the Greek heritage (having at best enriched it). It does not require us to investigate how Arabic philosophers and scientists fused and transformed the ideas from the different cultural traditions they inherited—a process that would implicate Europe itself with a much wider multicultural heritage derived from Arabic civilization. As we saw earlier, recognizing the Arabic heritage would have made the assimilation of the translated works within the religious context of medieval Europe much more difficult.

Let us now attempt to examine Arabic philosophy from the point of view of the multicultural approach we have been advocating. I would like to argue that the *falsafah* and *kalam* schools constitute different responses by Arabic thinkers to accommodate the philosophical and scientific traditions they had inherited from both the Greeks and Indians. The philosophical credentials of *kalam* are often questioned because of its concern with theological issues and its acceptance of a sacred scriptural tradition. However, if such an interpretation were accepted, we would also have to reject Augustine and Aquinas in Europe as having made contributions to philosophy. Equally, we would be led to reject much of what is historically taken to be Hindu and Buddhist philosophies. Given that we are prepared to see philosophy engage theological concerns in other cultures, we should also allow for the same objectives within *kalam*. Indeed if philosophy—*philosophia*—is the love of wisdom, then *kalam* has as much right to be deemed philosophy as *falsafah*.

Moreover, when we separate the debates of *falsafah* from *kalam*, we destroy any possibility of understanding many of the philosophical issues that confronted the proponents of *falsafah*, since these focused on their differences with the proponents of *kalam*, as well as their own theological interest in reconciling their monotheistic religious beliefs with Greek philosophy and science. By refusing to treat *kalam* thinkers as concerned with philosophical issues, we are forcing them into the procrustean bed of modern conceptions of philosophy—conceptions alien to both the traditions of *kalam* and *falsafah*. Worse, we are ignoring some of the key issues that divided *falsafah* and *kalam* and shaped the growth and evolution of both traditions in dialogical accommodation and conflict.[2]

Treating *kalam* as a philosophical tradition also enables us to appreciate its historical links to Indian atomic ideas and philosophical concerns. Many of the ideas characteristic of *kalam*—atomistic conceptions of the world, the search for union with God, the belief in intellectual illumination as a path to knowledge superior to empirical or rational methods, and the cultivation of the symbolic interpretation of religious texts—have parallels in Hindu and Buddhist, as well as Jain, philosophical traditions. What was unique about the *kalam* thinkers was conjoining such approaches with a commitment to their own monotheistic religious revelation, and adopting Greek logic to defend their positions. It was this synthesis of Indian and Greek philosophical approaches within the framework of a commitment to Quranic revelation that made it possible for al-Ghazali to develop his epistemological critique of Hellenic science that paved the way to modern science.

To understand the synthetic achievements of Arabic philosophy and science, let us begin with al-Kindi, often regarded as the first of the Arabic *falsafah*—he pioneered the integration of Greek philosophy and science into the monotheistic context of Islam. To treat him as a mere follower of Aristotle distorts the significance of his achievement. It is true that much of what he has to say takes off from Aristotle, but this is because he considered himself to be building on the heritage of the Greeks. Nevertheless, there are a number of crucial differences between Aristotelian and al-Kindian views. In his *History of Islamic Philosophy* Fakhry lists some of al-Kindi's differences with Aristotle—belief in creation ex nihilo, in the resurrection of the body, in the possibility of miracles, in the validity of prophetic revelation, in the Last Day when the world would be destroyed (Fakhry 2004, p. 70). Moreover these differences are intimately connected with al-Kindi's religious views, and they directed many of his key metaphysical, epistemological, and methodological views along paths not taken by Hellenic thought in general. This had significant consequences for the future development of Arabic philosophy—and subsequently, Western thought.

One result of the different religious and historical context in which al-Kindi approaches Greek thought is a transformation in what came to be perceived as the central problems of philosophy. The Hellenic thinkers were obsessed with the question of whether the forms of things were to be founded in particulars in the world or in some transcendental realm—the issue that divided Aristotelians and Platonists. However, both schools agreed that the perfect forms of mathematical ideas could not be embodied in the world. By contrast al-Kindi, like most Arab-Muslim philosophers who followed him, hardly sees this problem—in fact he does not take it up as a serious issue. Given his spiritual and theological orientation that assumes the existence of an omnipotent Creator of the universe—one whose power is assumed to be unlimited and therefore unconstrained by the limits faced by Plato's powerful Demiurge working on recalcitrant material to forge a cosmos, or Aristotle's Unmoved Mover engaged in contemplating his own thinking rather than the universe of imperfection and change—al-Kindi's God made the world "so that I can be known." According to the Quran he also made the universe a system of signs—*ayat*—to be read in order that He would come to be known (Nasr 1993, p. 462). Within such a conception of the created cosmos as a revelation of an omnipotent power, it would have appeared sacrilegious, as we saw earlier, to raise the question whether God could embody perfect mathematical forms in the cosmic order. For al-Kindi the perfection of the seen universe reveals the perfection of the Creator who cannot be seen—it is the visible sign of the invisible God, and its unity reflects the unity of God.

Consequently, when Greek philosophy entered the Islamic world, it was transformed, and came to be seen differently as a result of becoming naturalized into a completely distinct spiritual and eschatological context. As a result al-Kindi approached philosophy by combining the views of Plato and Aristotle in a fashion not possible within the Hellenic context. His works had the effect of making Arabic philosophers approach both Plato and Aristotle as together constituting a single unified tradition. Thus it is often noted that the Arabic Peripatetic tradition from al-Kindi to al-Farabi and Ibn Sina (Avicenna), with the possible exception of the Aristotelian purist Averroes, is really a neo-Platonic Aristotelianism—a contradiction within the Hellenistic framework but supremely natural within the Islamic context. This is not because Arabic philosophers were not aware of the differences between Plato and Aristotle—it was simply that these differences were minor compared with the basic problems they faced in attempting to reconcile Greek thought with Quranic revelation. Indeed al-Kindi's general approach to Greek philosophy was hardly ever questioned until the time of Averroes, near the end of the Arabic *falsafah* tradition, and after it had begun to collapse under the *kalam*-inspired critique of al-Ghazali. By that time the *falsafah* tradition itself had accomplished its greatest work.

Instead of the issue of reconciling Plato and Aristotle, a new problem arises in the Islamic world—how to combine Greek rationalism with the divine revelation of the Quran. At the time the Arabic thinkers took up this problem it had already been confronted by the other monotheistic communities that had preceded them, namely the Jews and the Christians. The influential synthesis of Proclus and Porphyry was an attempt to address this problem, and it was through this tradition that Arabic scholars entered Greek philosophy. Fakhry writes:

> The Greek work whose impact was most decisive on Arab philosophical thought was not, as might be expected, Aristotle's greatest venture into the realms of speculative thought, i.e. the *Metaphysica*, which had, as we have seen, found its way into Arabic as early as the middle of the ninth century. It was rather the *Theologia Aristotelis*, an alleged Aristotelian compilation whose Greek author is unknown. It was translated into Arabic for al-Kindi, the first purely philosophical Arab writer, around the same time as the *Metaphysica*, by a Syrian Christian, Abd al-Masih b. Naimah, of Emessa. The historical value of this work is considerable; it has been described as the epitome of Greek philosophy as it strove in Hellenistic times to blend into

a whole all the elements generated during the period of greater creativity. (Fakhry 2004, p. 21)

For Arabic *falsafah* concerned with harmonizing Greek philosophy with the conception of a monotheistic God, the *Theologia Aristotelis* was a great boon. It viewed the cosmos as an emanation of a single principle—identified with Plato's form of the Good—and made it possible for Arabic philosophers to assume that this principle was, in fact, their God who was also deemed to be both a Divine Unity and the source of all creation. The emanation theory was developed by al-Kindi and al-Farabi. It culminated with Ibn Sina (Avicenna), who argued that it could be demonstrated that there was only one uncaused Necessary Being—God—and that all other beings were the result of a series of necessary emanations that descended in a sequence, following the Ptolemaic model of the universe, from the stellar spheres through the different planetary spheres to the center of the universe where the earth is located.[3]

The emanation theory was contested by nearly all the theological schools. Not only did it seem to border on pantheism, and to violate the scriptural view that God created the world ex nihilo, but it also threatened the view that God was omnipotent. It did this by implying that the universe could not be other than what it is since, being a necessary emanation of God, it would not have been possible for God to create a different universe. As we saw earlier, these schools adopted a metaphysics of atomistic occasionalism— the world was made of instantaneous point atoms that appeared and disappeared without causally influencing one another. Only God existed as the sole and primary cause of events in the world, and the world was continuously generated ex nihilo by God's Will making all things be. Even the properties of the atoms—or *accidents,* as *kalam* evocatively described them—were assigned by God at the point of their appearance. This meant that by changing his mind God could create a different universe at any point in time. The world is ever at the mercy of God, and only the habits of God sustain the order we see in it.

Where did this atomic conception come from? Was it invented de novo by the Arabic scholars? *Kalam* atomism, so different from Hellenic atomism, is likely to have been derived from Buddhist views that maintained a similar conception of instantaneous point atoms. Moreover, atomism was a dominant view in Indian philosophy, since nearly all schools, whether Hindu, Buddhist, or Jain, accepted atomic conceptions of reality. These schools were quite influential at the time the Arab-Muslims emerged as the dominant power after their conquest of the Sassanian Empire, and their views must have been discussed at the centers of learning at Jundishapur, where scholars from the Hellenic and Indian worlds met. Arabic theological thinkers might have found such views attractive because Indian atomic views were closely linked with religious traditions such as Hinduism, Buddhism, and Jainism—and did not have the association with atheism they had in the Hellenistic world. Hence the Indian tradition of atomism is more likely to have been seen as acceptable by Arabic theologians than the Hellenic. Moreover, we know that historically *kalam* displayed unrelenting antipathy to Greek philosophy. This conclusion is supported by Fakhry, who argues that differences in *kalam* and Greek atomic views suggest that the former were inspired by Indian rather than Greek ideas:

> With hardly a single exception, the Muslim theologians accepted the atomic view of matter, space, and time, and built upon it an elaborate theological edifice over which God presided as an absolute sovereign. We shall have occasion to consider this atomic theory later, but it is noteworthy that some of its important divergences from Greek antecedents, such as the atomic nature of time, space and accidents, the perishability of atoms and accidents, appear to reflect an Indian influence. The two Buddhist sects of Vaibhashika and Sautrantika, the two Brahmin sects of Nyaya and Vaishashika, as well as the Jaina sect, had evolved by the fifth century an atomic theory, apparently independent of the Greek, in which the atomic character of matter, time and space was set forth and the perishable nature of the world resulting from their composition was emphasized. (Fakhry 2004, p. 35)

Moreover the association with religious thought also produced other similarities between Indian and Arabic atomism—*kalam* atomism did not deny the existence of minds independent of matter, belief in an afterlife, the need for spiritual cultivation, or the possibility of communion with a transcendent reality.

Thus, if *falsafah* can be seen as attempting to naturalize the Greek philosophical views into the spiritual horizon of the Quranic revelation, then *kalam* can be seen as naturalizing Indian atomic ideas into the same spiritual horizon. In the process we also see *falsafah* as accommodating the notion of knowledge based on reason into the Islamic universe, while *kalam* tried to do the same with the Indian concept of knowledge based on intuition. Initially *falsafah* tended to dominate creative philosophical thinking even though *kalam* developed in parallel with it. However, after the destruction of the Greek conception of rational knowledge by al-Ghazali, the Arabic world turned away from rational science toward intellectual illumination. Paradoxically, as we have seen, al-Ghazali accomplished this task by using the instrument of reason to subvert reason—essentially by showing that the intellect could not establish the first principles of any science.[4]

Standing apart from *falsafah* and *kalam* is the iconoclastic figure of al-Razi (Rhazes). He was one of the few influential thinkers in the Islamic world to reject outright the concept of revelation and the role of prophets as mediators between God and man. He argued that since God had given humans reason to guide them to the truth, prophecy was superfluous and, even worse, obnoxious since it caused much bloodshed by making people war against one another. Moreover, he believed in reincarnation, opposed the killing of animals for food, and believed that animals also had transmigrating souls. For these reasons he was considered to be an infidel by his contemporaries—someone who had come under the influence of Hinduism (Fakhry 2004, pp. 97–106).

Al-Razi is a seminal figure in science for two reasons. His medical corpus became canonical in European medicine after it was translated in the thirteenth century. His studies incorporated both the Greek and Indian medical knowledge available to him in his time (Goonatilake 1984, p. 51).[5] As a result of their transmission to Europe and their influence on the medical schools there until the beginning of the seventeenth century, they came to have a deep influence on shaping the context of development of modern medicine.

Moreover, al-Razi also contributed to the development of chemistry where, unique for his time in the Arabic world, he espoused atomic views. He is often considered to be a pioneer of modern chemistry because of the rational and nonmystical approach he made to the study of chemical phenomena. Since the emergence of science in the modern period was shaped by atomic ideas, it is possible that, given al-Razi's medical stature in Europe during the medieval period and his great impact on alchemical studies, he was also an influence on the emergence of atomic ideas in modern chemistry. Given his close interest in Indian medical works, which are also conditioned by the pervasive atomic conceptions found in Indian philosophical and religious texts, he may also be seen as indirectly transmitting Indian atomic ideas to Europe.

It is evident that we cannot treat the Arabic contribution to the philosophy and science of medieval Europe during the Renaissance as mere transmission of enriched Hellenic and Hellenistic thought. Being the heirs of Greek and Indian philosophical and metaphysical ideas, their views developed through the intense debates that divided *falsafah* and *kalam*. In applying the systematic deductive logic of the Greeks to defend its unique atomic views, inspired by Indian thought, *kalam* drew on both Greek and Indian heritage. Hence, even within *kalam* there was the meeting and fusion of Indian and Hellenic thought—a synthesis that was the unique creation of Arabic philosophy that went beyond both the traditions that inspired it.

The same may even be said of *falsafah*. Although it is considered to be inspired by only Greek thought, we have to remember that the neo-Platonic tradition that inspired it is also likely to have been conditioned in part by

Indian philosophical ideas. The emanantist doctrines of the neo-Platonist Plotinus, which *kalam* proponents suspect as introducing pantheistic ideas, is likely to have been influenced by Indian religious and philosophical ideas. Regarding Plotinus, Rawlinson writes:

> Plotinus, the founder of the neo-Platonic School, was so anxious to be instructed in Indian philosophy that he accompanied the expedition against [Shapur], King of Persia, in AD 242, in the hope that this might bring him into contact with someone who could help him. The resemblance between neo-Platonism and the Vedanta and yoga systems are very close. The absorption of the individual into the world is described by Plotinus in words which have a typically Indian ring ... neo-Platonism also has many points of contact with Buddhism, especially in enjoining the abstention from sacrifices and animal food. (Rawlinson 1975, pp. 435–436)

The most intriguing result of the Arabic attempt to naturalize Greek and Indian philosophical traditions into their spiritual universe ruled by an omnipotent deity is that it led in two directions that came to be integrated within Europe. First, it led Arabic *falsafah* to assume that the world obeyed a mathematical realist order, since an omnipotent God could, and would, embody the perfection of mathematics in the real world that served as the Creator's second revelation to humanity. Second, it led *kalam* to suppose that the world could not be constrained by regularities that were necessary, since this would place limits on the power of God. The regularities in the world were the habits of God—they could become otherwise if God changed his habits. Hence the only way to know these regularities is to look at the world. Empiricism became the orientation for those concerned with developing natural knowledge—though *kalam* thinkers themselves did not see natural knowledge as important.

Moreover, when Europe inherited the Alhazen optical theory it also inherited a theory that seemed to be both mathematical and realist in orientation and whose first principles could be, and had been, discovered by observation and experimentation. In many ways it seems reasonable to suppose that the union of mathematical realism and empiricism in Europe associated with modern science combined two streams of philosophical thinking that developed separately in the Arabic world as it adapted Greek and Indian traditions of thinking into its spiritual universe.

Apart from their synthetic and integrative achievements in philosophy—which they transmitted to Europe—Arabic scholars also made important contributions to mathematics by integrating the Indian and Persian algebraic approaches in mathematics with the geometric approach of the Greeks. The greatest exponent of algebra was al-Khwarizmi (c. 780–850 CE), who introduced the Indian decimal place system with zero to the Arabic world.

By contrast the greatest exponent of geometry and the deductive method was Thabit Ibn Qurra (836–901 CE), who is considered to belong to the Hellenic tradition. By the time of Omar Khayyam (1048–1131 CE) and al-Kashi (1380–1429 CE) these two traditions had merged together (Joseph 2000, p. 304).

Omar's great work—apart from his studies in the parallel postulate of Euclid that had also attracted the attention of Thabit Ibn Qurra and Alhazen—was on the theory of proportions. He recognized that Euclid's theory of proportions in his *Elements* had both an arithmetical and a geometrical aspect. However, Euclid had only considered rational numbers in his theory of proportions—or ratios. Omar extended Euclid's approach to include positive irrational numbers such as the square root of 2 and π. He did this by using geometric methods to solve algebraic equations. Referring to this accomplishment Joseph writes:

> [The Arabic thinkers] bring together two strands of mathematical thinking—the geometric approach which had been carefully cultivated by the Greeks, and the algebraic/algorithmic method which had been used to such effect by the Babylonians, Indians and Chinese .... By devising an efficient system of classifying equations, the Arab mathematicians, starting with al-Khwarizmi reduced all equations to three main types. For each type they offered solutions as well as a geometric rationale, thereby laying the foundation of modern algebra. Thus were the *ahl al-jabr* (the "algebra people"), in Thabit's words, and the "geometry people" brought together. (Joseph 2000, p. 328)

The Arabic achievement of bringing algebra under geometric methods was repeated from the other side when Descartes (1596–1650 CE), with his analytic geometry, brought geometry under algebraic methods—essentially by representing lines and curves by algebraic functions. Descartes describes his project as inspired by a search for the "true method" for the sciences. He writes in his *On Method:*

> Among the different branches of philosophy, I had in my younger days to a certain extent studied logic; and in those of mathematics, geometrical analysis and algebra—three arts or sciences which seemed as though they ought to contribute something to the design I had in view.[6]

He proceeds to argue that in order to correct the limitations of the geometric approach we would have to borrow from the best in algebra and vice versa. Descartes came to find the key by recognizing that every problem in geometry is concerned with proportions between lines—precisely the insight that Omar Khayyam previously had. This enabled him to develop algebraic methods to solve geometrical problems—thereby doing in reverse what had been accomplished by Omar Khayyam when he extended geometric methods into algebra.

Since the Arabic achievements in philosophy, mathematics, and medicine involved a synthesis of Hellenic and Indian ideas, they cannot be treated one-sidedly as elaborations of Greek thought. They have to be seen as attempts to fuse together elements from multiple cultural traditions. They are similar in scope and range to the attempts to integrate Greek and Christian thought by Augustine or Aquinas, or Buddhist and Confucian thought by Zhu Xi. It is precisely the confrontation of systems of thought very different from each other brought together in the Islamic universe that stimulated Arabic thinkers. They created a novel tradition through their synthesis of philosophical, scientific, medical, and mathematical ideas from many cultures, especially the Hellenic and Indian traditions. By passing this synthesis to Europe—one that cannot be reduced to just Greek or just Indian thought—they shaped the development of medieval European thought and ultimately modern science and philosophy.

Chapter 11

Universal Mathematical Laws in a Mechanical Universe

After the seminal studies of Needham, historians of science have generally conceded that Chinese discoveries in technology from the beginning of the Han dynasty to about 1500 far outpaced in ingenuity and sophistication those developed in most other civilizations. However, Needham also argued that the cosmological and philosophical orientation of Chinese thought precluded the development of modern science in China. According to Needham, Chinese correlative cosmology, which explained phenomena in terms of each part of the cosmos accommodating itself to changes in other parts to promote universal harmony, obstructed Chinese thought, by preventing the development of the experimental approach for studying nature, and the notion of universal natural laws governing the universe. Needham saw the main obstacle to the emergence of the concept of universal natural law among the Chinese to be their nontheistic spiritual orientation. This made it difficult for them to conceive the universe as regulated by laws, since such a conception can only arise in the context of belief in a being that sets these laws.

However, the concept of universal laws did not arise in Greek science either in spite of their theistic belief in divine beings regulating the cosmos. Thus belief in divine beings need not give rise to the notion of universal laws—it requires, as Needham would affirm, the belief in a monotheistic God capable of laying down laws for the universe as a whole. There was no such God in Greek thought and therefore nothing like a universal law. Neither the Platonic doctrine that heavenly bodies only perform circular motions, nor the Archimedean laws of the lever, pulleys, or hydrostatics have universal application. Rather they are contextual laws applying to specific domains of phenomena—the planets and stars, levers, pulleys, bodies immersed in fluids, and so on. By contrast the modern laws discovered in the seventeenth century apply to all material bodies throughout the universe—Newton's laws of motion, the laws of conservation of mass and momentum, and the law of gravitation have universal application.

There is also another reason why the notion of universal laws could not have arisen in Greek science. The tradition did not see one universe but two—a universe of change in the sublunary sphere below the moon, and a universe of permanence in the superlunary heavens, except for circular motion. Given this cosmological orientation it would not have been possible for Hellenic scientists to assume the existence of universal laws, since the realm of the heavens not only was made of a different kind of matter (Aristotle's *quintessence*) but also obeyed a different set of regularities. The bifurcated cosmos of Hellenic science cannot be obedient to a single set of universal natural laws.

The notion of universal laws regulating the cosmos only emerged in the modern era in Europe. According to Needham the earliest clear concept of a law of nature is to be found in the seventeenth century:

> [T]he first use of the expression "law of Nature" in its scientific sense occurs in 1665 in the first volume of the early scientific journal called *The Philosophical Transactions of the Royal Society of London,* and within thirty years it had become a commonplace. (Ronan 1978, vol. 1, p. 289)[1]

He goes on to note that Copernicus never mentions "laws" but symmetries and harmonies when referring to the motion of the planets; neither does Galileo refer to laws. Only in Descartes, according to Needham, does the concept become fully articulated when he speaks of the "law God has put into Nature," and discusses his laws of momentum as mechanical laws that imposed themselves upon all bodies in the universe (Ronan 1978, vol. 1, p. 290).

Moreover, the modern view also assumes, along with the notion of universal laws, the additional notion that these laws of nature are mathematical regularities. Indeed it was Galileo who stressed that the book of nature is written in the language of mathematics. As we have seen the Greeks never developed the faith that ideal mathematical relations could be found in the world—the concept of nature as perfectly obedient to mathematical principles was not a part of Greek science. Thus Ptolemy set himself only the task of saving the phenomena; and Euclid considered his geometrical theorems to apply without qualification only to ideal mathematical objects, but thought that real objects in the world cannot be perfect triangles or perfect circles. In their optical studies Ptolemy and Euclid were satisfied with a mathematical theory that was physically highly implausible—the extra-mission theory that Alhazen refuted. Even the great Archimedes worked with ideal levers and ideal pulleys. The view of nature as embodying ideal mathematical order only became possible, as we have argued, with Alhazen's optical theory which provided the first credible exemplar of the notion that mathematical forms in all their perfection could be applied to natural phenomena. Nevertheless, even this theory did not postulate universal laws of nature since it confined its applicability to optics, though it did make credible the mathematical realist view that natural phenomena could be regulated by perfect mathematical laws.

However, for mathematical realism to combine with the notion of universal laws that were laid down by God there had to be another change. The Greek concept of cause as necessarily linked to effect had to be broken in order to make God free to create natural laws by an act of divine free will. It was al-Ghazali's destructive dialectics that opened the door to this change. We have seen al-Ghazali argue that God, being omnipotent, could have created a world different from the one we now perceive—a point denied by the Aristotelians and their supporters like Averroes. He made the regularities of the world the habits of God—a change that made it possible to imagine that these regularities applied universally across the cosmos.

The significance of the transformation in thinking wrought by al-Ghazali becomes evident when we take into account Duhem's claim that the crucial event which led to the birth of modern science was the decree by the Church in Paris against the Aristotelian view that there could be only one possible world. He writes:

> If we had to assign a date to the birth of modern Science, without doubt we would choose this year 1277 when the Bishop of Paris solemnly proclaimed that several Worlds could exist, and that the entirety of the heavenly spheres could, without contradiction, be animated by rectilinear motion.[2]

According to Duhem this decree liberated scholars from the out-of-date teachings of Aristotle and Averroes and made it possible for them to speculate on alternative ways of interpreting the world. The decree asserts the omnipotence of God and God's freedom to make a world obedient to a different order of causes and effects than what we actually find. Thus, heavenly bodies do not *necessarily* have to perform circular motions—they do so only because God has commanded them so. According to Duhem the decree is the key event that led to the modern concept of natural laws that have to be empirically discovered because they are not necessary truths.

If Duhem is correct, then al-Ghazali should also get some credit for this edict. The decree actually institutes in Europe precisely the claim made by *kalam* in the Islamic world, and does this shortly after the translation of Arabic works in Europe had begun, and in the context of Averroes' defense of Aristotelian science against al-Ghazali's attack. We could argue that the decree in Paris compelled obedience to the idea that al-Ghazali fought so hard to entrench—that the regularities in nature are products of the Will of God, and not self-evident or necessary principles required by reason and discoverable by its application.[3]

Thus, given the intimate links between the intellectual currents in thirteenth-century Europe and those in the Arabic world we must assume that Duhem's claim that the key event which led to the birth of modern science was a decree in 1277 by a bishop in Paris to be overly simple. Rather the decree in Paris has to be seen as the ultimate outcome and response to the influence of the Averroist movement in Europe. Moreover, Averroism itself was an answer to al-Ghazali's critique of *falsafah*. Consequently, when the decree attacks Averroism and Aristotelianism to defend the notion that there are no necessary truths which constrain God's creation of the world, it is also giving indirect support to al-Ghazali's critique of *falsafah*. The decree of 1277 reflects precisely the demands of the dominant theological school in the Arabic world—demands equally relevant in the European Christian context confronting Hellenic philosophy.

Consequently the emergence of the concept of universal natural law that has to be discovered by empirical methods, and that Duhem sees as triggered by the Paris decree of 1277, must now also be seen as conditioned by the debates between *falsafah* and *kalam*. Moreover, the mathematical realist conception of natural laws can also be traced to Arabic science—especially to the seminal work of Alhazen. However, the combination of mathematics and universality into universal mathematical laws of nature only happened in Europe in the modern era. In a sense it integrated the *falsafah* demand for mathematical realism in science with the *kalam* view that the habits of God make the laws of nature.

There is another dimension to the modern notion of laws of nature that we have not yet taken into account. Apart from being both universal and mathematical they are also mechanical laws—laws that apply to mechanical systems. This distinctive mechanical vision of nature is not to be found either in Alhazen's optics or in Ghazalian philosophy. I would like to argue that the fusion of the concept of a universal mathematical law with the notion of a mechanical law is the outcome of the meeting in Europe of Arabic philosophy and science with Chinese mechanical discoveries. It was in the Renaissance era that this confluence first began, especially in the thirteenth and fourteenth centuries when parts of Western Europe in Andalusia were under Arab-Islamic rule, and parts of Eastern Europe in Russia were under Mongol rule. Lying in the middle, the territories that

later came to pioneer the development of modern science became the ter-
rain into which technologies and ideas flowed from the Chinese east and
the Arabic west. These territories were the areas where the *kalam*-inspired
notion of universal laws and *falsafah*-inspired mathematical realism fused
with the mechanical vision inspired by Chinese technologies.

Is it an exaggeration to claim that the flow of Chinese technologies
transformed European sensibilities in the mechanical direction? A mere
listing of the technological flood should make us reconsider any reservations we might have about its potential impact—the square-pallet chain
pump, metallurgical blowing-engines operated by water-power, the wheel-
barrow, the sailing-carriage, the wagon-mill, the cross-bow, the technique
of deep drilling, the so-called Carden suspension, the segmented arch-
bridge, canal lock-gates, numerous inventions in ship-construction (including water-tight compartments and aerodynamically efficient sails),
gunpowder, the magnetic compass for navigation, paper, printing and
porcelain (Needham 1954, pp. 240–241).

Moreover, in case we continue to have doubts about the influence these
technologies wrought on European thinkers we can turn to the writings of
Francis Bacon, who was sufficiently impressed by them to make them the
basis for his call to create a new mechanical order of science:

> It is well to observe the force and virtue and consequences of discoveries.
> These are to be seen nowhere more conspicuously that in those three which
> were unknown to the ancients, and of which the origin, though recent, is
> obscure and inglorious; namely, printing, gunpowder, and the magnet. For
> these three have changed the whole face and state of things throughout the
> world, the first in literature, the second in warfare, the third in navigation;
> whence have followed innumerable changes; insomuch that no empire, no
> sect, no star, seems to have exerted greater power and influence in human
> affairs than these *mechanical discoveries.* [My emphasis][4] (Quoted in
> Needham 1954, p. 19)

This passage, reflecting the view of one of the leading lights of the new science in Europe, and referring to three discoveries unknown to the ancients
he considers to have had a greater influence on Europe than empires or
religions, should make evident the profound impact of Chinese technologies on European consciousness—for all the inventions Bacon alludes to
are Chinese discoveries. The methodological movement that developed in
Europe with Bacon, Descartes, and others, and that led to modern science,
can be seen as the attempt to capture the "essence" of these technologies in
a new mechanical vision.

However, the mechanical vision of the universe may also have had a
theoretical influence from the Chinese over and above the impact of their

technologies in transforming European consciousness. One of the great discoveries often seen to have influenced the mechanical turn in the seventeenth century was Harvey's discovery of the circulation of blood and the mechanical explanation of the heart as a pump that controlled this process.[4] However, Harvey published his discovery in 1628 more than a century after Europeans had established themselves in China. This raises the question whether the discovery of the circulation of blood, and the mechanical conception of the heart as a pump associated with it, may not have been influenced by Chinese ideas.

Indeed the circulation of blood was a central tenet of Chinese medical theory. By the second century BCE the doctrine had become incorporated into the *Yellow Emperor's Classic of Internal Medicine* (Veith 1966)—a text that established the fundamental principles of Chinese medicine and influenced all subsequent Chinese medical development until modern times. It was written as a dialogue between the emperor and his minister and dealt with all questions of health and arts of healing. It soon became a canonical text for Chinese medical practitioners. One of its cardinal teachings is that blood circulates through the body, although its description of the circulatory system is somewhat vague. Nevertheless, it does make a close link with the circulation of blood, the heart, and the pulse by saying "the heart is in accord with the pulse" (Veith 1966, Book III, chap. 10), so that taking the pulse became established as a fundamental diagnostic tool of Chinese medical practice.

Ultimately the theory led the Chinese to see the heart as a pump so that Chinese doctors were trained to see its action by looking at the model of a system of bellows pumping liquid through bamboo tubes. This surprisingly mechanical understanding of cardiac activity could easily have been transmitted to Europe by Jesuit missionary scientists in China, concerned with collecting knowledge of medical pharmacopoeia, and expanding medical knowledge, as one of the strange ideas of the Chinese only to become subsequently incorporated into a major discovery of early modern science (Lu and Needham 1980, pp. 32–33).

There is, however, another influence that could have complemented the Chinese impact by lending further credence to the notion that blood flows through a circulatory system. This is the discovery in the thirteenth century by the Egyptian physician Ibn al-Nafis of the circulation of blood from the heart to the lungs and back to the heart again, that is, what has now come to be known as the lesser (or pulmonary) circulation of blood. His writings are included as parts of his commentary on Ibn Sina's *Canon of Medicine* in which al-Nafis criticizes Galen's teachings concerning the heart and blood system (Rapson 1982, p. 44). In the second century Galen had argued that the liver produced blood. He taught that the blood reaching the right side of the heart traveled through tiny invisible pores in the cardiac septum to the left side, where it became mixed with air to create spirit before being distributed to the rest of the body. Ibn al-Nafis correctly identified that blood from the right side of the heart first went to the lungs, whose bronchi he described, where it mixed with air before returning to the left side of the heart. He even showed that the coronary arteries served to supply blood to the cardiac musculature.

Ibn al-Nafis' writings were translated and published in Venice in 1547. It is significant that shortly thereafter we find other Europeans, now credited with anticipating Harvey, also proposing circulatory theories concerning blood—Servetus in 1553, Colombo in 1559, and Cesalpino in 1571. It is therefore very likely that European medical practitioners may have been led to question Galen's views not only because they had been aware from reports since the early sixteenth century that the Chinese believed blood circulated the body, but also by coming to know of the discovery of the lesser circulation of blood by Ibn al-Nafis. The evidence adduced by al-Nafis of the lesser circulation made more convincing the possibility of a larger circulation. By the time Harvey published *De Motu Cordis* in 1628 the notions of lesser and larger circulation of blood from the Arabic and Chinese medical traditions had become fused into one integrated theory.

The discovery of the circulation of blood also had another impact of a more philosophical kind. By showing that the heart worked like a mechanical pump, a metaphor quite close to the Chinese idea that it pumped blood the way bellows pump air, it made even more credible the emerging mechanical philosophy against its Aristotelian opponents. For Aristotelians believed that the heart was the seat of emotional activities—a view that has become naturalized into everyday speech when we speak of a person's "heartache" or "lion-hearted" character (Rapson 1982, p. 23). The new mechanical view of the heart became one more weapon in the arsenal deployed against Aristotelian doctrines—over and above new discoveries in astronomical and dynamical theories.

Thus it becomes possible to understand how the mechanical vision of nature developed in Europe as a result of the impact of two kinds of influences from China. First, it came to be inspired by the slew of mechanical technologies in the medieval period that transformed sensibilities in a mechanical direction, as made evident in Bacon's remarks quoted earlier. Second, the impact came directly through the notion of the heart as an organ pumping blood through the body that provided a powerful metaphor to consolidate this sensibility. This is not the whole answer of course—the mechanical vision took off only after Newton when the notion that the universe as a whole was a mechanical system became credible. However, what is being suggested is that the Chinese technologies, and the Chinese theories of the role of the heart in the economy of the body, paved the way to this mechanical vision of the universe.

The hypothesis we are proposing raises another question. If the new science in Europe was so profoundly influenced by the discoveries of China why did these not have the same impact on the Chinese. Why were the Chinese not equally inspired by their discoveries to be led to a mechanical vision of the universe? We have already suggested earlier that this is because the discoveries took place over a long time in China, and entered Chinese culture one by one so that they could be integrated into the organic materialist view that dominated Chinese thought. However, the impact in Europe was far greater because these discoveries had to be absorbed over a brief period in historical terms.

But another answer to this question has been offered by the sinologist A. C. Graham. He argues that in China practical and theoretical concerns were kept separate, and not brought together as they were in Europe. This prevented the development of a theoretical vision designed to integrate the technological discoveries made in China. He argues that we fail to recognize this today because we tend to link technology and science together. He writes:

> It has been our habit to think of science and technology as from the first intimately connected, and gradually progressing side by side as causal thinking comes to prevail over correlative thinking of primitive magic and Mediaeval proto-science. Our bias inclines us therefore, either to think of China as on the verge of modern science but somehow (for reasons which it would be of great interest to discover) failing to reach it, or else to be suspicious of claims to Chinese priorities in technology. (Graham 1989, p. 315)

Graham finds both alternatives unacceptable—neither was China on the verge of modern science because of its technological advances nor can we question its priority in technological discoveries simply because it did not achieve modern science. He explains how it is possible for the Chinese to be both technologically progressive and yet scientifically regressive:

> It must be taken for granted that in China concern for the practically useful stimulated causal thinking in technology as strongly as in the West, and contributed as much or more to material wellbeing until it was outstripped in the last few centuries. But to suppose that this would be bringing China nearer to modern science assumes an obsolete conception of science as developing by continuous progress in rationality. We now think in terms rather of a Scientific Revolution about A.D. 1600 the "discovery of how to discover," the quite sudden integration of the idea of explaining all natural phenomena by mathematized laws of nature testable by controlled experiment. (Graham 1989, p. 317)

Graham then goes on to argue that the Chinese separated their thinking into two parts—when it came to theoretical and philosophical discourse they engaged in correlative thinking and adopted a correlative cosmology. However, when it came to practical matters where the pursuit of utility was important they engaged in causal thinking.

Graham's explanation hardly provides a satisfactory answer to the question addressed. First, it does not tell us why the Chinese adopted a causal approach when they engaged in practical pursuits. Second, it fails to explain why their correlative philosophical orientation did not obstruct their causal approach to dealing with practical matters, including the design, construction, and implementation of a large number of sophisticated technologies over an extended period of time.

It is far more reasonable to suppose, contra what historians of Chinese science such as Graham and Bodde assume, that it was precisely the correlative cosmology of the Chinese that made them so successful in the arena of technological discoveries. Such an explanation becomes even more credible when we compare the notion of cause in Aristotelian philosophy and the notion of cause in Chinese correlative thinking—the former inhibits and the latter liberates the technological experimentalism essential for innovation in mechanical engineering.

For Aristotle understanding a natural phenomenon involves explaining how it is brought about, as we have seen, by the action of the four types of causes—the final, material, formal, and efficient. However, these causes themselves ultimately can be grounded on the more fundamental concepts of matter and form that Aristotle uses to analyze all processes of change. Jones uses the example of a growing acorn to illustrate how Aristotle would analyze the process through which it becomes an oak. If we think of the oak tree as planted then the final cause is the purpose for which it is being grown; the material cause is the soil, water, and other things needed to nurture its growth; the formal cause is the form of the tree that exists potentially in the acorn; and the efficient cause is the person who planted the acorn. In the case of a naturally growing oak tree Jones admits that it is problematic to give final and efficient causes (Jones 1969, pp. 223–225).

He continues to add that on this Aristotelian view the acorn can be seen to carry the potentiality of being an oak tree, and the oak is the actuality of this potentiality. The environment in which the acorn grows merely provides the nurturing medium in which the oak is formed by realizing the potential in the acorn. Thus the giant oak tree is an outcome of something essential already present in the acorn, albeit in a manner not yet actualized.

This is in sharp contrast to the Chinese correlative conception of causality, which considers the oak to be what it is, not by virtue of some essence within it, but by virtue of the context in which it grew. It is its relations with other things that make it what it is. This is not to deny that the seed has a role but it is only one factor—among many others—that has to be taken into account. From the Chinese point of view the oak is explained correlatively in terms of the larger context in which it grew, whereas Aristotelians explain it in terms of essential properties within the original seed that produced it. Both are causal explanations of the oak but one looks to correlated causes in the context outside and treats the initial seed as but a facilitating factor; the other looks to necessary causes within the seed and treats the context as a facilitating factor.[5]

I want to suggest that when we deal with mechanical systems the correlative causal approach is far more likely to promote creativity and innovation than the causal essentialist approach, since the functioning of a part in a machine is not determined by the material out of which it is made (it could be made of wood, iron, or glass), or its form (which could vary considerably so long as it performs its function in its location), or its purpose (for the same part in different locations may have different purposes); only the efficient cause that puts it in the location in the machinery may have to be taken into account. This means that the Chinese can pursue the understanding of a thing in terms of how it harmonizes (or fits in) with other parts of a system. It is the harmony of the whole, and the way each part fits into the larger system, that explains why it behaves as it does. Explaining a thing in terms of how it correlates and works in harmony with the whole is, for the Chinese, explaining it. It appeals not to the essence of a thing, or essential causes within it, but how it meshes into the whole system—that is, in terms of its correlation with other things.

According to Needham this holistic orientation is reflected in the philosophy expounded by the neo-Confucian philosopher Zhu Xi (1130–1200) who saw nature as regulated by "principles of organization" that he terms *li*. Needham considers the notion of *li* to be close, but not quite identical, with the modern conception of natural law because it also includes a conception where the parts are seen as fitting into the whole (Needham 1956, p. 558). This conception of law is more in accord with the Chinese organic materialist philosophy of nature, in which the regularities of nature are seen as dependent on the relationships between things in nature (Needham 1956, pp. 518–583). Indeed another way of interpreting Needham is that he sees *li* as embodying contextual laws of nature—a notion quite different from the modern notion of universal laws of nature.

Moreover, an understanding of a thing, not in terms of its essence but in terms of its context, would also promote a tinkering orientation to a system as a whole, in order to see how the behavior of a part would change when we change other parts in the system. This trial and error process is precisely what is involved in the experimental method where we change the context of the object we study in order to understand how its behavior is thereby affected. Such an experimental approach would be more difficult to take seriously if we supposed that a thing behaves as it does by virtue of an essence within it.

I would like to suggest that Bacon's experimental method is a systematization of the technological experimentalism that produced the Chinese mechanical technologies transmitted to Europe. What Bacon did, as the Chinese did not, was to systematize this tinkering orientation by creating new contexts never found in nature—"torturing nature," as he put it—in order to discover those invariant correlations that occurred in all experimental contexts. Thus it was the inspiration provided by the impact of Chinese technology, and its experimental tinkering orientation approach to mechanical discovery, that led Bacon to the "discovery of how to discover."

However, we cannot ignore the intermediate stage in the turn to Bacon's active experimental method pioneered by the Arabic scientists who, such as Alhazen and Razi, influenced the experimental orientations of Roger Bacon and Grosseteste and, according to Hess, even that of Galileo and Kepler (Hess 1995, p. 66). We have seen that Kant emphasizes the importance of the change brought about by Bacon's active experimental method in contrast to the passive collection of facts that had been the methodological orientation of Greek science.[6] However, what he fails to see is the transitional stage of the passive experimental method introduced by Arabic scientists and practiced in Chinese science. In the passive experimental method the experimenter works within the contexts that arise in nature, but imposes variations on these contexts in order to study the laws that regulate nature in various contexts. This goes beyond the passive method of observation of the Greeks, but falls short of the method of active experimentation pioneered by Bacon that demands we violate natural contexts to determine the laws that regulate nature in all contexts.

Finally it is to be noted that the Chinese correlative notion of the cause-effect relationship is also close to Hume's view of the cause-effect nexus. We have already seen that according to Hume the regularities in nature we observe, when we think that a cause A invariably leads to an effect B, are not regularities that follow necessarily but only contingently. Neither observation nor logic can demonstrate that B necessarily follows from A— all that we can claim is the observed regular succession of B given A, that is, all that we can claim is that occurrences of A are *correlated* with occurrences of B. It follows that scientific laws merely correlate events A with events B, and what we consider to be causal laws in which a cause A necessarily leads to an effect B are really correlative laws showing that events A are regularly correlated with events B.

Hence, Hume may be said to have shown that the notion that an effect B is necessarily linked to a cause A is not a part of modern science—it is a metaphysical idea. Moreover, this metaphysical idea was, as we have seen, an integral part of Greek science that was effectively criticized by al-Ghazali. Thus, far from being an obstacle to the development of Chinese technology or the rise of modern scientific ideas, the correlative vision made possible the Chinese technological discoveries that played such a key role in the rise of modern science and the emergence of the mechanical philosophy of nature. Moreover, after Hume this correlative conception has continued to be an important way in which scientists understand the cause-effect nexus.

However, the various elements derived from Arabic and Chinese science and technology only came together to create modern science in Europe. Within Europe the experimental approach to discovering correlative regularities in nature and the mechanical sensibility became united with a mathematical realist orientation to nature seen as obedient to universal laws. In the process ideas from Chinese science and technology fused with those of Arabic science and technology to generate a new science. That Europe should be the geographical arena of this new synthesis should not be surprising— it was sandwiched between two extensive and powerful civilizations that held territories in the west in the Iberian Peninsula under Islamic rule, and territories in the east in Russia under Mongol rule. Caught between the hammer and the anvil, and under the intense threat it faced in both directions, Europe responded creatively by drawing on the resources of the civilizations on its borders to create the key elements of modern science.

Chapter 12

Fusing Solar and Stellar Cosmologies

When Copernicus set the earth in motion around the sun in 1543 he gave two motions to the terrestrial sphere—one a rotational motion around its axis every twenty-four hours, and the other a revolutionary motion around the sun nearly every 365¼ days. Assuming such a motion, the heavens above would appear to exhibit two kinds of regularities. First, the daily rotation would give the appearance that the sun, moon, planets, and stars revolved in a circle around an axis defined by the north and south poles of the earth and, in the Northern Hemisphere, this would give the appearance that they revolved around the pole star, located more or less vertically above the North Pole. Second, the annual revolution of the earth around the sun would give the appearance that the sun was drifting along a path across the zodiac, since the line joining the observer on earth to the sun would project onto different constellations of stars as the earth orbited the sun. Moreover, since the earth's axis was not vertically inclined to the plane of revolution, but tilted by approximately 23 degrees, the path of the sun would move in a circle that was inclined by the same amount to the equator projected onto the sky. This circle is called the ecliptic.

As a result of these two apparent motions—one of heavenly bodies around the pole star each day, and the other of the sun along the ecliptic every year—ancient astronomers had the choice of making their measurements relative to the pole star or relative to the sun as it raced around the ecliptic. Following the Egyptian and Mesopotamian pioneers in astronomy the Greek, Indian, and Arabic astronomers defined positions of heavenly bodies in relation to the sun's path across the sky. By contrast the Chinese defined astronomical positions relative to the pole star. This had profound consequences for the development of the two traditions, and the Copernican Revolution—by revealing both the motion of the sun and the motion around the pole star to arise from the dual revolutionary and rotational motions of the earth—brought the two traditions of astronomy together. This change had significant consequences not only for the development of astronomy in the seventeenth century, but also for science in general.[1]

How did this confluence of traditions come about? One significant factor that led to the change was the influence of Jesuits who arrived as missionaries in China in the sixteenth century. By the end of that century the Jesuits had established themselves in the Chinese imperial court as astronomers largely because they carried with them better techniques for both calendar calculations and the prediction of eclipses. This brought them into intimate contact with Chinese astronomical ideas, which they communicated back to Europe on a regular basis. Hence it can be said that the Jesuits established a corridor of communication between European and Chinese astronomical traditions so that the future development of astronomical ideas in Europe occurred in a context that could not ignore Chinese views of the heavens—and, as we see later, did not.

These Chinese views were so different from those held by Europeans that the Jesuits considered them incredible. By the same token the Chinese probably considered the European theories equally bizarre. This is evident from the letter of the Jesuit astronomer Matteo Ricci who wrote home in 1595 that the Chinese believed in a number of "absurdities"—as he called them—including the following:

1. The earth is flat and square, and that the sky is a round canopy; they did not conceive the possibility of the antipodes.
2. There is only one sky (and not ten skies). It is empty (and not solid). The stars move in a void (instead of being attached to the firmament).
3. As they do not know what air is, where we say there is air (between the spheres) they affirm that there is a void.
4. By adding metal and wood, and omitting air, they count five elements (instead of four)—metal, wood, fire, water and earth. Still worse, they make out that these elements are engendered the one by the other; and it may be imagined with how little foundation they teach it, but as it is a doctrine handed down from their ancient sages, no one dares to attack it.
5. For eclipses of the sun, they give a very good reason, namely that the moon, when it is near the sun, diminishes its light.
6. During the night, the sun hides under a mountain which is situated near the earth.[2]

What is striking is that Ricci—who was well versed in the European astronomical theories of his time (precisely the reason he was allowed entry into the highest Chinese circles)—should have made a list in which half the absurdities he refers to have become a part of modern astronomy. Given current science we would be prepared to acknowledge that it is absurd of the Chinese to think the earth is square shaped, that the antipodes do not exist, or that the sun hides behind a mountain at night. However, we would agree with the Chinese and think it absurd that Ricci would believe that there are ten skies against the Chinese view that there is only one; that he would imagine the stars to be fixed to a solid firmament and not float in void space as the Chinese affirmed; that the air extended to the stars and was not merely a small envelop around the earth; and that elements like wood, water, or earth could not, given our present knowledge of physics, be transformed into each other as the Chinese believed. Hence we would agree with the Chinese on statements two, three, and four and with Ricci on statements one and six. Statement five is somewhat ambiguous—did the Chinese think the moon diminished the light from the sun by blocking it or by having some other sort of influence upon it? The letter does not explain, but Ricci, who must have known the cause of solar eclipses, thinks the Chinese reason to be good. However, what is significant is the very different judgment we would now make about the so-called absurdities Ricci writes about. What divides us from Ricci is the modern scientific revolution, in which many "absurdities" from Chinese and European science (as seen by each other) were combined into a new synthesis that came to supplant the two traditions Ricci compared.

In order to appreciate how this confluence of the two traditions occurred let us examine the European and Chinese cosmological views at the time Ricci was communicating his Chinese "absurdities" back to Europe. It was a time a little over fifty years after Copernicus had proposed his theory, but this theory had made little headway among astronomers. It was also forty years before the indictment of Galileo and the banning of the theory by the Catholic Church. At the time of Ricci's communication the Copernican theory was perceived as hardly threatening to religious orthodoxy because it appeared so physically implausible—it could not explain why bodies fell if the earth was not at the center of universe; it could not account for the absence of gale force winds and raging seas despite its assumption that the earth rotated at incredible speed; and the absence of stellar parallax as the earth completed its vast orbit around the sun was another item of crucial evidence against the theory. The Copernican theory was acknowledged to be a good mathematical hypothesis able to save the phenomena, but a poor alternative to the Ptolemaic model that accomplished the same task with more physical plausibility.

Thus the Ptolemaic model of the universe continued to hold sway among most of the astronomers of Europe. This model posited that the earth was at the center of the universe and the seven planets—the bodies that drifted against the background of the stars—were fixed to heavenly celestial spheres that, by their perpetual rotation, carried them across the heavens. The order of the planets was as follows: beginning with the moon nearest the earth we move outward to the spheres of Mercury, Venus, Sun, Mars, Jupiter, and Saturn. Beyond the sphere of Saturn was the sphere of the fixed stars—the *primum mobile*. The spheres that carried the heavenly bodies were themselves considered to be made of a crystalline substance—the *quintessence*—so transparent that we are able to see the stars through all the seven spheres onto which the lower planets were attached. During the Renaissance era Christian writers had felt justified in adding two more spheres—an outermost, invisible, and immobile Sphere of the Empyrium, which they considered to be the abode of the angels, and an aqueous crystalline sphere of water between it and the stellar sphere. This brought the number of spheres to ten—the ten skies that Ricci referred to (Lindberg 1992, pp. 248–250).

Moreover, there were other properties assigned to this cosmos. All the space between the spheres was filled with air because Aristotle had maintained that nature abhors a vacuum. The objects above the lunar sphere underwent no change except for the rotation of the spheres. Change did not occur because heavenly bodies were made of the refined substance *quintessence*. By contrast below the lunar sphere we encounter a world of change because all sublunary things were made of the elements earth, air, fire, and water, which ceaselessly combine, separate, and recombine in innumerable ways.

Thus the medieval cosmos is a bifurcated cosmos separated into two distinct realms—the sublunary and the superlunary—subject to two different kinds of order. It is also a spatially finite cosmos even though there was some speculation as to what lay beyond the last sphere. It was a cosmos without any empty space—a cosmic *plenum*—in which every part was filled with air extending up to the stellar sphere. Even more important it was a stable cosmos—once created by God it did not change (except below the moon). Even on the earth things remained stable in the larger context because the continents and seas did not change after their creation (except that some religious dogmatists allowed those changes required to accommodate the biblical flood).

In contrast to their European counterparts the Chinese had three different theories of the universe, all of which were developed in the Han dynasty. In many ways the situation paralleled that of optics before the Alhazen revolution, when the discipline was divided between three schools adopting the Platonic, Aristotelian, or atomist theories of light phenomena, each capable of resolving some problems but not others.[3]

The earliest of these models is generally considered to belong to the *gai tian* tradition developed in the first century BCE. Its originator is unknown but the original text called *Zhou bi* is available. The theory holds the sky and the earth to be made of two plates running parallel to each other with a space between them. The sky is a round disc that rises up as a dome in the center which covers the earth that is a square, which also rises up in the center like an inverted plate. The sky disc is fixed to an axis at its center located near the pole star, about which it rotates carrying the other heavenly bodies with it. The sun also moves with the sky disc but at the same time moves to and from the central axis at different times of the year, being furthest away from the axis in winter. Also the sun's light is assumed to reach only a finite distance so that regions beyond its reach remain dark. The virtues of the model lay in the fact that it could explain day and night successions by the axial rotation of the sky disc and the variations of the seasons by the movement of the sun toward and away from the central axis. This was probably the theory that Ricci referred to when he wrote of the absurd Chinese belief that the earth was a square and the sky a round canopy.

The second Chinese theory was the *hun tian* theory, first proposed in about 100 BCE but elaborated more fully by Zhang Heng (78–139 CE) in the early second century. Zhang describes it as follows:

> The heavens are like a hen's egg; the earth is like the yolk of the egg, and lies alone in the center. Heaven is large and earth small. Inside the lower part of the heavens there is water. The heavens are supported by *qi*, the earth floats on the waters.[4]

Zhang then proceeds to present the heavens in the form of a celestial globe as we do today. He also pioneered the construction of an equatorial armillary sphere in China that had a system of rings corresponding to the great circles of the celestial sphere. In the middle ran a central tube used to line up stars and planets so that Zhang could make more precise star maps than his predecessors.

He also pioneered a method for mapping the stars seen in the celestial sphere onto a grid system drawn on a plane surface by inventing conformal projection. This technique is what has now come to be described as the Mercator projection, discovered in Europe in 1568. The advantage of conformal projection is that it preserves shapes of constellations and directions between stars on a sphere when these are represented on a flat surface, although at the price of distorting sizes of constellations. The projection can be made by inserting a transparent globe, on which the constellations are mapped, into a cylinder and turning on a light inside the globe. The images projected onto the cylinder yield a conformal projection of the objects represented on the sphere. Of course a flat map can be made by copying the projections on a piece of paper wrapped around the cylinder and then straightening it (Teresi 2002, p. 152).

However, the most interesting and philosophically sophisticated of the Chinese theories was the *xuan ye* theory, which assumed that the earth and other heavenly bodies were floating in an infinite empty space. This theory is described by Ko Hung (283–343 CE) in the early fourth century, and he formulates the views of the original supporters of this model as follows:[5]

> They said that the heavens were empty and void of substance. When we look up at it we can see that it is immensely high and far away, having no bounds. The (human) eye is (as it were) color-blind, and the pupil short-sighted; this is why the heavens appear deeply blue. It is like seeing yellow mountains sideways at a great distance, for then they appear all blue. Or when we gaze down into a valley a thousand fathoms deep it seems somber and black. But the blue (of the mountains) is not a true color; nor is the dark color (of the valley) really its own.
>
> The sun, moon, and the company of stars float (freely) in the empty space, moving or standing still. All are condensed vapor. Thus the seven luminaries [planets] sometimes appear and sometimes disappear, sometimes move forward and sometimes retrograde, seeming to follow each a different series of regularities; their advances and recessions are not the same. It is because they are not rooted (to any basis) or tied together that their movements can vary so much. Among the heavenly bodies the pole star always keeps its place, and the Great Bear never sinks below the horizon as do other stars. (Ronan 1981, pp. 86–87)

This theory came to be endorsed by the members of influential neo-Confucian orthodoxy, especially Chang Tsai (Ronan 1981, p. 88). Chang also argued that, adopting the *yin-yang* cosmology, the earth was made of pure *yin* that had condensed into a solid at the center of the universe, and that the heavens were made of pure *yang*. The heavens revolved in an anti-clockwise direction around the pole star. The movement of the stars was due to a floating, rushing *qi* that carried them in their orbits around the pole star. However, it was also known that unlike the stars, the sun, the moon, and the planets (the seven luminaries of Chinese astronomy) sometimes moved in an opposite direction. This was explained by postulating that these bodies, being nearer the earth, were subject to drag by the viscous resistance of the *qi* of the earth. According to Needham these views were current among Chinese astronomers at the time the Jesuits arrived in China (Ronan 1981, p. 88).

The European and Chinese views were so different at the time that it is informative to compare them. For the European astronomers the center of revolution of the heavens is the earth; for the Chinese it is the pole star. From a qualitative point of view of saving the appearances the two models are equally acceptable, since the Chinese theory also holds the earth to be at the center of the universe. However, by fixing the center of revolution in the pole star there is no temptation for the Chinese astronomers to treat the sun, the moon, and the planets as performing circular orbits around the earth.

Indeed since these bodies move erratically in relation to the regular circular motion of the stars it also becomes possible to see them as freely floating in space. Moreover, both the theory that the stars and planets are driven by the push of a *qi* wind and the explanation that the more complex motions of the planets are due to the viscous drag of the *qi* of the earth, show the Chinese theory proposing qualitative physical explanations of heavenly motions. This is in sharp contrast to Ptolemy's rejection of physical explanations—Ptolemy recognized that although his mathematical model could "save the phenomena," it was physically implausible.

Another crucial difference between the European and Chinese astronomical views is the absence in China of any sharp distinction between the substance that made heavenly bodies and bodies on earth. Although the earth is pure *yin* and the heavens are *yang*, these properties are only, as is well recognized, seen in relation to each other. For the notion of *yin-yang* is also used to explain processes on the earth. Hence the Chinese cosmos is not a bifurcated cosmos like the European cosmos was at that time. Indeed phenomena connected with the earth could affect heavenly bodies and were used by the Chinese to explain their behavior—such as the explanation of the planets' retrograde motion by reference to the drag of the earth's *qi*.

Perhaps the most significant difference between the two astronomical traditions was the Chinese belief, associated with the *xuan ye* theory, that space is empty and infinite. This was later to be seen as one of the most important changes brought about by the new cosmology of modern science. The notion of an infinite empty universe was not even possible to entertain within the Aristotelian worldview of the Jesuits, in which infinity could only be an attribute of God, and nature was held to abhor the vacuum. Alexandre Koyré has argued that the most profound mutation in thought triggered by the seventeenth-century scientific revolution involved the change from a closed world to an infinite universe (Koyré 1957). If the change also pointed the way from a Greek Aristotelian view of the cosmos to that of the Chinese *xuan ye* theory.

Another important difference between the Chinese and European views is the notion that heavenly bodies have a history and are subject to continual change. According to the Chinese view the sun, moon, and stars condensed out of vaporous *qi*. The *qi* that fills all of empty space, pushes the stars in their revolutions, and retards the planets is also the material out of which they were originally formed. This is strikingly close to modern views that stars form out of dispersed matter in empty space and that the solar system condensed from dispersed gas in space. It offers a picture of a universe in perpetual transformation so different from the immutable heavens assumed by the European astronomers in the sixteenth century— and even well into the eighteenth.

Learning that the Chinese viewed the realm of stars as having a fluid changing nature Christopher Scheiner writes in 1625:

> The peoples of China [he said] have never taught in any of their innumerable and flourishing academies that the heavens are solid; or so we may conclude from their printed books, dating from all times during the past two millennia. Hence one can see that the theory of a liquid heavens is really very ancient, and could easily be demonstrated; moreover one must not despise the fact that it seems to have been given as a natural enlightenment to all peoples. The Chinese are so attached to it that they consider the contrary opinion (a multiplicity of solid celestial spheres) perfectly absurd, as those inform us who have returned from among them. (Quoted in Needham 1958, p. 6)[6]

Given the numerous differences between the views of Chinese and European astronomers when they first met in the sixteenth century, and given that views were to develop in Europe that paralleled Chinese views in the seventeenth century, is it reasonable to suppose that European astronomers were not influenced by these Chinese ideas? Such a conclusion would hardly be credible. After all we have to keep in mind that the Jesuits were members of one of the most learned societies in Europe; they were deeply interested in the new sciences emerging in Europe, and many of them were accepted in the Chinese court precisely on the grounds of their astronomical skill in calendar calculations, and at a time when questions pertaining to astronomy were becoming controversial in Europe, and alternative models of the cosmos were becoming increasingly investigated. Under such circumstances can we reasonably assume that the dominant thematic ideas of Chinese astronomy were not of intense concern to European astronomers and that they would have remained indifferent to them?

Moreover, there is abundant literature describing the intense debates that took place in China between Jesuit and Chinese astronomers, with each group trying to win for themselves the task of defining the astronomical agenda for the Chinese world. These Jesuit astronomers were also in constant communication with their church superiors and counterparts in Europe.[7] However, it is hardly likely that we are going to find some European astronomer referring to a Chinese astronomer to support views such as the infinity of space, the voidness of space, or the condensation of planets and stars out of vaporous material. What is more likely, and what we do find, is European astronomers and scientists referring back to ideas within their own tradition to support notions similar to those in Chinese science, which were communicated back by the Jesuits.

Alexandre Koyré, in his book *From the Closed World to the Infinite Universe*, argues that the notion of infinite empty space entered modern astronomy through Nicholas of Cusa, and treats him as a forerunner of Bruno, Copernicus, Kepler, and Descartes (Koyré 1957, pp. 19–20). Indeed the same idea can be traced back to the Greek atomists—Lucretius, Diogenes Laertius, and Epicurus. However, does this mean that the real reason this idea came to be taken seriously in the seventeenth century is the influence of these thinkers who had been on the margins of the dominant tradition of natural philosophy at that time? May not the fact that these ideas were the central notions of the dominant tradition of Chinese astronomy be the key factor in promoting greater attention to them in Europe? Is it not possible that once they had come to be seriously entertained European scientists foraged within their own tradition for precedents to give them historical legitimacy and authority? Clearly the answers we give to these questions would determine the direction taken by our history of modern science. If we treat the ideas in Europe as a case of independent discoveries that can be traced back to European antecedents we would be reinscribing a Eurocentric history. However, if we see them as the products of the influence of Chinese ideas we open the door to a dialogical perspective.

At this fork in the road I would like to appeal to the thematic transmission criterion proposed earlier to decide whether we ought to take the case for dialogical influence seriously. Applying this principle we can now ask the following question: Did a new corridor of communication open up between China and Europe at the time we are concerned with? Yes, it opened in 1514 when the first Portuguese, Jorge Alvarez, arrived in China. It is easily conceivable that Portuguese captains and sailors would have become acquainted with basic Chinese astronomical ideas after this—especially given the Portuguese interest, wherever they went, of acquiring maps of the seas from the cultures they came in contact with and hiring local pilots to navigate them through foreign waters. In showing the Portuguese how they navigated by the stars it is possible that their Chinese guides would also have transmitted some basic and rudimentary notions of the dominant Chinese astronomical tradition. This may not have been as nuanced and sophisticated as what Chinese astronomers could have offered, but it would at least have been as much as the average Portuguese captain or sailor could have taught their Chinese counterparts. The Portuguese could have taught the Chinese about the existence of crystalline heavenly spheres, the existence of air that filled space up to the stars, and so on. Equally, no great sophistication is needed to transmit the notion of empty space, or that heavenly bodies condense from vaporous substance.

However, the corridor of communication that opened in 1514 widened with the setting up of the Macau depot in 1557, and communication through it became significantly enhanced with the arrival of Matteo Ricci in Macao in 1582.[8] Ricci was to become one of the most talented students of Chinese astronomy, directly appointed by the emperor to work with Chinese astronomers and well versed in European astronomical ideas and theories at the same time. His written communiqués to Europe on Chinese astronomy have to be treated as studies of comparative astronomy by a master of the two traditions.

Thus, the opening of the corridor was also accompanied by a deep interest on the part of Europeans to understand Chinese astronomy. The fact that the Jesuits made their mark in China because of their better skills in calendar calculations and prediction of eclipses—a matter of crucial significance to the imperial court given the Confucian belief that heavenly events were portents of the future state of the empire—ensured that the Jesuits would pay careful attention to Chinese astronomy. Moreover, since most of them were also trained scientists with a natural interest in scientific matters, their studies were carefully examined by colleagues in Europe with whom they maintained constant communication. Thus, not only did a new corridor of communication open between China and Europe in the sixteenth century, but it was also accompanied by an intense interest in Chinese scientific ideas.

Let us now compare the dominant themes of Chinese science against those in Europe. The Chinese believed in infinite empty space, in a cosmos of universal change and flux, in heavenly bodies as condensed out of primordial fluid *qi*, and other so-called "absurdities" for the Jesuits. These views were, moreover, strongly supported by the neo-Confucian orthodoxy that advised the rulers of China. Were these themes dominant in Europe during the sixteenth century? We have to give a negative answer since such notions came to be taken seriously only after the revolutionary changes wrought by the modern astronomical and scientific revolution. Since the new ideas in Europe came shortly after a corridor of communication opened between China and Europe, following intense interest in Chinese views, and were similar to the dominant themes of Chinese thought at the time, it is reasonable to suppose that they were the outcome of the impact of Chinese science on the European creators of modern science.

Moreover, the impact of Chinese astronomical ideas can be considered to have continued even after the end of the seventeenth century and into the eighteenth century. It is interesting to note that Zhu Xi, whose thought dominated neo-Confucian philosophy at the time the Jesuits entered China, and whose views Leibniz sought to integrate with European ideas, had proposed an evolutionary theory of the universe that in many ways is similar to Kant's view centuries later. This has been noticed by Sun Xiaochun:

> He (Zhu Xi) said the universe began as permeating *qi* (ether), which separates into *yin* (cold, negative) and *yang* (warm, positive) *qi*. When the *yin* and *yang qi* grind against each other, they produce some dross, and the *qi* begins to condense around the cores of dross to form celestial bodies …. This theory is quite similar to Kant's nebulae hypothesis of the origin of the universe, but it was formulated seven centuries earlier. (Sun 2000, p. 424)

Given these similarities one can ask if Kant could have been unaware that in the debates between Chinese and European astronomers, some Chinese held heavenly bodies to continually form and dissipate through the condensation and evaporation of *qi*.

Indeed once we accept that Chinese ideas were transmitted to Europe through the corridor of communication opened by the Portuguese voyages of discovery to the East at the end of the fifteenth century, we also have to take seriously the notion that the influence may have been far more extended than hitherto suspected. We have seen that the Chinese did not subscribe to the notion of a bifurcated cosmos like Greek, Arabic, and European astronomers. This notion had led Europeans to believe that no significant changes, apart from circular motions, occurred above the lunar sphere in the heavens. Unimpeded by such dogma the Chinese were able to identify a vast range of astronomical phenomena that made it possible to demolish the conception of an unchanging heavens in modern astronomy.

The Chinese knew that the sun was not a perfect sphere and had studied sunspots since the time of Liu Hsiang in 28 BCE. By 1638 they had described in detail 112 sunspots in their official records, as well as noting them in other published documents. However, in Europe these were first noticed by Galileo only in 1610—and constituted one source of evidence against the Aristotelian conception of a static heavens (Ronan 1981, p. 211). Moreover, it is important to note that European astronomers in China would already have been aware of the importance that the Chinese attached to recording sunspot activity. Prior to Galileo and after the arrival of Europeans in China there were numerous such records—in 1511, 1518, 1562, 1566, 1567, 1569, 1590, 1597, and even in 1613, when Galileo published his letter on sunspots.

The Chinese astronomers also kept complete records of comets, meteors, and meteorites, and the orbits of some forty comets before 1500 are available today only in Chinese records. Their observation of meteors and meteorites go back as far as 687 BCE. A *Leonid Shower* reported by the Chinese in 931 CE gives such precise descriptions that it is now possible to use it to make a statistical analysis of the observations and confirm its periods of recurrence (Ronan 1981, p. 210).

In the case of novae and supernovae the Chinese records go back to 1300 BCE—their first record of a nova is found on an oracle-bone. Of the four supernovae documented in our local galaxy in historical records, the earliest two are given by the Chinese in 1010 CE and 1054 CE—the latter being the year of origin of the Crab Nebula. Only after a corridor of communication opened to the Chinese world do we find European astronomers recording such events as astronomical phenomena—Tycho Brahe studied the nova of 1572 and Kepler that of another in 1604 (Ronan 1981, p. 205).

Although comets, novae, and supernovae were observed outside China, they were not treated in Greek, Arabic, or medieval European science as astronomical phenomena. Often they were explained away as atmospheric events. In medieval Europe, for example, comets were believed to be hot, dry exhalations that had ascended into the heavens from the earth. Hence they were included along with phenomena like lightning, thunder, shooting stars, and rainbows.

Yet such observation of sunspots, comets, and novae—a whole class of phenomena unknown to European astronomy prior to the arrival of the Portuguese in China in the early years of the sixteenth century—were to constitute evidence against the Aristotelian and Ptolemaic cosmologies in the seventeenth century. The theme of a changing heavens in which new stars, comets, and sunspots made their surprising appearances entered European astronomy only after its contact with Chinese astronomy, where it constituted a dominant theme. Is it not reasonable to assume that the awareness of these Chinese views made European astronomers receptive to the study of sunspots, comets, and novae by enabling them to recognize these phenomena as the celestial events that they really were? (Lindberg 1992, p. 252).[9]

How, it may be asked, could such an influence have been transmitted? After all, it could be said, Tycho Brahe studied the famous nova of 1572 and recognized it to be a superlunary event a decade before Ricci arrived in China. Hence it is unlikely that he could have known about Chinese views since no European with astronomical interests had reached China prior to that time.

However, this argument ignores the fact that it does not require sophisticated training in astronomy to transmit ideas such as the existence of comets, novae, or sunspots.[10] Portuguese captains and sailors engaged in conversation with their Chinese counterparts about stars and planetary alignments for navigation, could easily have learned that the Chinese saw the heavens to be in continual transformation, in contrast to their conception of immutable celestial spheres. Such ideas could easily have been transmitted back to Europe and become seen as part of the Chinese "absurdities" that

Europeans had to tolerate. In the process European thinkers themselves would have been exposed to an alternative way of seeing the heavens. Consequently when a comet, a nova, or a sunspot appeared, it would have made European astronomers look at these more carefully without dismissing them as terrestrial exhalations of no concern to them. Hence, even if the Chinese ideas transmitted orally were not believed at first because of their seeming absurdity, they would nevertheless have become a part of the European imagination—and would later have come to constitute a part of the European world.

Consider other discoveries in the European world that simply happened to occur after the opening of the corridor to China and that paralleled discoveries already known to the Chinese. The equatorial mounting for telescopes used by Tycho Brahe, and now the standard mounting for such instruments, was the same mounting used by Chinese astronomers for their instruments. Also known to the Chinese was the use of equatorial coordinates in astronomy adopted by sixteenth-century European astronomers.[11] The equal angle projection introduced in Europe by Mercator in 1548, was known to the Chinese since the time of Zhang Heng in the second century, as we have seen. Moreover, all of these discoveries can be seen as natural extensions of the Chinese interest in the astronomy of the stars, in contrast to the focus on planetary astronomy within European, Arabic, and Indian traditions. For example, given the importance they attached to the pole star it was natural for the Chinese to adopt equatorial mounting and equatorial coordinates that took into account the rotation of the earth about its axis—although the Chinese projected this rotation to take place around the pole star.

It was also natural for them to produce the most detailed star maps and to discover the conformal projection that Mercator extended to produce a map of the spherical earth. Mercator's adoption of this projection, used by the Chinese to represent constellations seen on a celestial sphere, to represent land areas on the spherical earth, gave navigators a powerful instrument that simplified their task. It enabled them to navigate in a fixed direction on the globe of the earth by drawing a straight line on a conformal map that represents it.

One could also point to other discoveries made in Europe after contact with China, such as Harvey's discovery of the circulation of blood and Gilbert's discovery of the properties of magnets. What is important to note is that all these discoveries could have been easily transmitted from China, and they constitute the sort of knowledge that could be acquired by a European captain of a ship working on maps and stellar coordinates; by a doctor on a ship listening to his Chinese counterparts explaining an illness as due to poor circulation of blood; by a ship's guide speaking of the properties of a magnet and the lines of geomancy it followed, and so on.

However, what is clear is that even if we consider each one of these discoveries, taken in isolation, to have been an independent discovery of some European scientist or scholar, it begs credibility to assume that the whole slew of them—involving notions so alien to European conceptions of nature prior to the Chinese connection, and so close to dominant themes of Chinese thought and practice—were independent discoveries in Europe that happened to coincide with a time beginning more than fifty years after a corridor of communication was opened between China and Europe. It is more reasonable to assume that just as Chinese technology transformed European sensibilities after the Mongol Empire opened a corridor of communication between them in the thirteenth and fourteenth centuries, so did Chinese scientific ideas and techniques influence Europe when the Portuguese opened a new corridor through the oceans. Clearly when Copernicus set the world in motion to spin on its axis and to revolve around the sun, he made it possible for solar astronomical ideas, based on the study of the motions of heavenly bodies against the path of the sun, to fuse with the stellar astronomical ideas of the Chinese based on the study of heavenly motions against the pole star.

Chapter 13

The Wider Copernican Revolution

The Copernican Revolution is often considered to be the central event that triggered the Scientific Revolution of the seventeenth century, and it is linked to the names of Galileo, Kepler, Descartes, and Newton, who in turn used it as the fulcrum to give birth to modern science. The core idea of the Copernican Revolution is a distinctively European achievement, for the only person prior to Copernicus to entertain it seriously was the Greek Pythagorean astronomer Aristarchus.[1] No other civilization had ever independently considered the possibility of the earth as both rotating on its axis and revolving around the sun. The Arabic astronomers knew of the theory but only as one proposed by an ancient Greek, but no Arabic scientist considered developing it systematically because it was deemed physically implausible. In the fifth century CE the Indian astronomer Aryabhata had argued for the rotation of the earth but he did not consider it to also revolve around the sun. The Chinese astronomers had recognized a wobbly motion of the earth to account for certain astronomical observations but this was far from the Copernican theory.

Given that the Copernican theory developed out of Ptolemaic theory, and as a response to perceived inadequacies within it, it is often argued that the Copernican Revolution can be explained by appealing to internal European developments in astronomy. We have seen why this conclusion is unwarranted—the revolution itself could not have occurred without the contributions of Indian mathematics and Arabic critiques of the Ptolemaic theory on physical realist grounds. Moreover, the success of the theory also depended on shattering the concept of immutable crystalline heavenly spheres—since the theory itself made the earth move through a space deemed by Ptolemy to be part of the heavens—and much of the evidence that made this possible came from Chinese astronomy. Hence, even if the core Copernican idea of a rotating and revolving earth is purely European, a whole host of ancillary mathematical, theoretical, and empirical discoveries drawn from a diversity of cultures were needed to make this idea credible.

We have seen that Kuhn considers the Copernican Revolution to refer to two different changes—the narrow revolution in astronomical theory initiated by Copernicus, and the wider revolution associated with the science that came to completion with Newton.[2] According to Kuhn the Copernican theory created new problems for terrestrial physics. It could not explain why bodies fell to the earth if it was not at the center of the universe; it also created new problems for celestial physics—such as why the earth orbited the sun if it was not attached to any revolving celestial sphere. While dealing with these problems Galileo was led to study the motion of projectiles and make discoveries that subverted Aristotelian physics, and Kepler was led to his three laws of planetary motion. Newton finished the task of synthesizing the physics of Galileo with the astronomy of Kepler. We have already seen that the narrow Copernican Revolution was made possible through the impact of multicultural contributions. Let us now examine whether the wider revolution was not also made possible by drawing upon multicultural traditions of science.

It may be supposed that even if the narrow revolution of the Copernican theory required multicultural input the wider revolution may not have needed further input. The argument sounds extremely plausible given Kuhn's view that the achievements of Galileo, Kepler, and Newton were unanticipated consequences of accommodating astronomical and physical theory to the heliocentric vision. Indeed Kuhn does not even recognize the need for any dialogical explanation of the earlier narrow Copernican Revolution since he considers Greek science to have been on the threshold of modern science. Moreover, he also argues that it could have developed directly into modern science if contingent historical factors had not led to its decline in the European Dark Ages. For Kuhn the role of Arabic civilization was simply that of preserving, adding, and transmitting the Greek heritage to Europe. He writes:

> Therefore, from our present restricted viewpoint, Islamic civilization is important primarily because it preserved and proliferated the records of ancient Greek science for later European scholars. (Kuhn 1957, p. 102)

We will now proceed to examine carefully Kuhn's arguments for supposing that the Copernican Revolution, both in the narrow sense of the change made by Copernicus and the wider changes that led to Newton's synthesis, followed an internal logic designed to accommodate the hypothesis of a moving earth into astronomical and physical theory. Although much has been written since Kuhn's study on the Copernican Revolution, there has hardly been any attempt to show the immanent intellectual motives that drove the revolution in later studies. It may be suspected that Kuhn himself made such approaches difficult after his seminal work, *The Structure of Scientific Revolutions,* written five years later. It made the kind of "logicist" and intellectual approach adopted in his account of the Copernican Revolution less fashionable and opened the door to wider sociocultural approaches to the revolution.

But Kuhn's study still remains important because by doing a close reading of his historical narrative we can follow the Copernican Revolution in detail through its various actors. Moreover, since Kuhn first published his study there has grown an abundant literature on multicultural contributions to early modern science that makes it possible to identify those points where they should have entered his narrative. From this new perspective we find that Kuhn's history distorts the wider Copernican Revolution in two significant ways. First, in tracing the historical origins of various ideas he breaks the narrative at crucial points where multicultural influences enter the picture. Second, he grounds major influences on the wider (as well as the narrow) Copernican Revolution—the authority of Hermetic notions, the mathematical realist orientation of the pioneers of the new science, the atomic ideas that emerged in the seventeenth century, the mechanical vision associated with the changes that took place—on minor themes within the Greek tradition, even after European contact with cultures in which these existed as major themes. By such deforming fields imposed upon the historical narrative—albeit without deliberate intent—Kuhn appears to make a plausible case for a purely Eurocentric history of the Scientific Revolution that ushered in modern science.[3]

Let us now reread Kuhn's historical narrative of the Copernican Revolution by following it through, but including the multicultural impacts not perceived by him, in order to see the how the picture would change as a result. To make for completeness this study will trace Kuhn's historical construction through both the narrow and the wider Copernican Revolution, using some of the discoveries we have already made earlier regarding the multicultural impact on the narrow revolution.

Kuhn begins his account by examining what motivated Copernicus to make the revolutionary shift from the geocentric to a heliocentric theory of the planets. According to him the background for this was laid by medieval thought. Hellenistic scientists had been prepared to accept the bifurcation of astronomy and cosmology to a certain extent, but medieval scholastic thinkers were not prepared to go along with this separation. They saw Aristotle and Ptolemy—although separated by five centuries, and by the two distinct traditions of Hellenic and Hellenistic science—as proposing a single body of doctrines. Hence they saw the contradictions between Ptolemaic mathematical astronomy and Aristotelian cosmology as casting doubt upon the whole tradition of ancient science (Kuhn 1957, p. 105).

What Kuhn leaves out of this picture, however, is that this doubt itself came about precisely because of the failure of Arabic thinkers to reconcile the Aristotelian and Ptolemaic traditions. These attempts had led Arabic scientists to modify the Ptolemaic tradition and to question some of its fundamental assumptions, and their efforts in this direction were inherited by medieval European astronomers. We have seen that Alhazen, and later the Andalusian philosopher-scientists Averroes and al-Bitruji, followed by the Maragha School astronomers, all had strong reservations about Ptolemaic astronomy—especially in relation to its use of the equant. These criticisms had an impact on Copernicus and motivated his reconstruction of the astronomical tradition. However, we have also seen that the Arabic efforts at reconstructing Ptolemy's theory were not simply attempts to reconcile Aristotle and Ptolemy; they were also a demand for mathematical realism in astronomical theory rooted in their religious beliefs.

Their mathematical realist orientation also led them to raise another objection against Ptolemy's theory—it could not explain why heavy bodies fell to the earth. According to Aristotelian physics all heavy bodies tend to move toward the center of the universe and fall to the earth because the center of the earth was at the center of the universe. However, in order to create a mathematically adequate account of planetary motion Ptolemy had located the center of the universe slightly away from the center of the earth. This led to a conflict between Aristotle's physical theory explaining why bodies fell to the earth by virtue of trying to get to the center of the universe, and Ptolemy's mathematical model designed to give an account of the observed motion of the planets in terms of circular motions (inspired by Plato). This conflict between the physics of local motion and the mathematics of celestial phenomena did not bother Ptolemy, who saw mathematical astronomy as merely designed to "save the phenomena."

The Arabic scholars' commitment to mathematical realism also led them to criticize the Ptolemaic model of celestial spheres as physically unrealizable. For one thing these spheres intersected each other without shattering one another. According to the historian of Arabic science Seyyed Hossein Nasr, the physical realism of the Arabic scientists led them to "solidify" the abstract heavenly spheres of Ptolemy (Nasr 1968, p. 176). In his treatise on the constitution of the heavens, the mathematical astronomer Thabit Ibn Qurra conceived of the heavens as made of solid spheres that were filled with a compressible fluid between them. His attempt at a physically plausible mathematical model of the planets was studied by both Maimonides and Albertus Magnus, who included excerpts of his writings in their studies. These would clearly have influenced medieval European astronomers addressing the same problems.

Thus the demand for a mathematical realist astronomical theory was fairly widespread in the Arabic world. Many attempts were made over the whole period of Arabic astronomy to find a more adequate model than that given by Ptolemy. Nevertheless, despite their reservations, Arabic scientists did not give up the commitment to a geocentric universe—their concern was that any mathematical theory of the heavens should also be physically plausible at the same time. The Greek view of separating the pure realm of mathematics from the impure realm of the mundane world did not satisfy them. Moreover, in their search for alternative models, scientists such as al-Biruni even came to consider the possibility of the motion of the earth around the sun, as well as the possibility that the planets moved in elliptical rather than circular paths—both came to constitute central assumptions of the science that developed in the seventeenth century. By ignoring these Arabic explorations, their persistent demand for a physically plausible and mathematically adequate theory of the planets, and the models they developed to conform to this demand, we would distort the wider multicultural influences on Copernicus that led him to reject the Ptolemaic theory.

Kuhn also argues that the cosmological revolution to heliocentric theory made by Copernicus was only a small part of his achievement since the heliocentric idea was already ancient. According to Kuhn Copernicus' seminal contribution lay in demonstrating the mathematical cogency of the idea of a moving earth. However, Saliba, following studies by Neugebauer and Swerdlow, argues that the mathematical instruments used by Copernicus had already been developed by the Maragha School and that Copernicus merely changed the vector connecting the earth and the sun in the al-Shatir model. Combining the views of Kuhn and Saliba we might have to conclude that Copernicus was original neither as astronomer nor as mathematician!

However, this would be to ignore the radical nature of the synthesis he wrought—Copernicus used the mathematical instrument and theory of the geocentric Maragha School to construct a heliocentric theory of the planetary system. It was this feat of synthesis that made his work original—not the component parts that went into it. In this sense Copernicus can be seen as drawing on the results of a long movement in Arabic astronomy critical of Ptolemaic theory, but not of geocentric theories in general, to construct a model of the universe that subverted geocentrism altogether. Such an understanding of his achievement makes him a dialogical figure who combined ideas of the European Aristarchus with those of the Arabic astronomer al-Shatir. He is far more revolutionary than either Kuhn or Saliba consider him to be.

According to Kuhn another influence that shaped the Copernican Revolution—the wider revolution—was the impetus theory first proposed by medieval thinkers. This theory is significant for the rise of modern science because it removed some of the most serious objections to the Earth's motion that were rooted in Aristotelian physics. The impetus theory began as an attempt to deal with the inadequacies in accounts of motion provided by Aristotle, and constituted the first stage for the inertia theory that played a crucial role in the rise of modern science. The law of inertia is Newton's first law of motion—namely, that a body would move in a straight line with uniform velocity unless forces acted upon it. By contrast Aristotle had maintained that a moving body would come to rest if no force were to act on it. This is the commonsense view—everyday observation seems to suggest that a moving stone, say, would come to rest unless it was kept in motion by a force supplied externally, or fall to its natural resting place at the center of the earth if unsupported.

For medieval thinkers the Aristotelian account of motion created problems. Adopting the theory seems to force us to conclude that when a stone is hurled as a projectile it should, immediately after leaving the hand, fall directly to the ground because there is no external push on it. The problem had been recognized by Aristotle who sought to get around it by arguing that the moving stone cuts through the air threatening to create a vacuum. Since a vacuum is impossible the surrounding air would rush in to prevent this, thereby propelling the stone forward. This explanation was unconvincing—it did not explain why heavy stones traveled further than, for example, cotton balls of the same size. Given the same push the lighter balls should go further but experience shows they do not. Attempts to deal with this objection by claiming that cotton balls, being lighter, felt the resistance of the air more severely were unconvincing—they required one to assume that the air was responsible for both retarding and propelling the ball at the same time.

The impetus theory assumes that when a projectile is set in motion it acquires a "motive force" that becomes stored in it, and gets dissipated as it moves through the air. According to this theory, proposed by John Philoponus in the sixth century and developed by Jean Buridan in the fourteenth century, a projectile would move rapidly at first, then more slowly as its impetus gets depleted, and fall to the ground when the impetus is exhausted. By the end of the fourteenth century, according to Kuhn, impetus dynamics in one of various forms had replaced Aristotelian dynamics. It was taught in Padua at the time Copernicus studied there, and was also learnt by Galileo from his teacher Bonamico. Given its importance as a stage for the law of inertia, and given that Copernicus himself appeals to a version of it to explain why the rotating earth does not cause the bodies on its surface to be thrown about, Kuhn argues that this medieval achievement played an important role in the emergence and consolidation of modern science (Kuhn 1957, pp. 115–123).

What is significant about Kuhn's history of the development of the impetus theory is that, although he refers to the early work of Philoponus and the medieval proposal of Buridan, he leaves out completely the Arabic contributions to the impetus theory. Yet it was the Arabic scholars who took up Philoponus' theory and developed it to the point from which European scholastics were to take off. Following Philoponus' criticism of Aristotelian dynamics Ibn Sina (Avicenna) developed the notion of *mayl* (translated into Latin as *inclinatio*) to account for the motion of projectiles. His theory was further developed by Abu 'l-Barakat al-Baghdadi and known to the scholastics. Moreover, the concept of momentum was discussed by Alhazen in his *Optical Thesaurus,* and given the pervasive influence of this work on European science until the times of Galileo and Kepler, it could hardly have been ignored by them (Nasr 1976, p. 139).

The Andalusian philosopher Ibn Bajjah (Avempace) was to create what became known in European scholastic circles as "Avempacian dynamics." This was proposed as an alternative to Aristotelian dynamics. Aristotle had argued that for a moving body, using modern formulations of his notion in the language of algebra, if we let $V =$ its velocity, $P =$ the motive power acting on it, and $M =$ the resisting power of the medium, then $V = P/M$. However, this had the unfortunate implication that when $M = 0$ (i.e., in a vacuum where there is no resistance) V would become infinite. This did not bother Aristotle because he considered that a vacuum was not possible. This was also to become one more argument in favor of the impossibility of a vacuum for Aristotelians, since if a vacuum was possible then bodies would travel in it at an infinite velocity—an absurd conclusion. However, given that the Arab-Muslims maintained that God could create a vacuum (and some Andalusian philosophers had even argued that beyond the last heavenly sphere there was a vacuum) this Aristotelian solution was unacceptable. Consequently Ibn Bajjah (Avempace) argued that we should have $V = P - M$, so that in a vacuum the velocity of a body would still be finite. This Avempacian dynamics was known to the medieval scholastics, having been discussed by Averroes in his commentary on Book IV of Aristotle's *Physics.* Moreover, when Galileo attempted to deal with Aristotelian objections to the possibility of a vacuum, he also turned to Ibn Bajjah's dynamical theory (Nasr 1976, p. 139).

However, the impetus theory was not studied only by Arabic and European scholars; Indian scholars also used a similar theory when they addressed the problem of motion. According to Ronan:

> What the Indian view suggested was that when a body first experiences the force that sets it moving, the very application of this force imparts a quality,

vega or impetus, which causes the body to continue to move in the same way. When a body meets an obstacle it either comes to rest or continues moving, but more slowly; how slow depends on how much the obstacle has neutralized the impetus: complete neutralization results, of course, in a stoppage.

This doctrine of impetus was a notable contribution to thoughts and explanations about the motion of bodies. In the West the Aristotelian doctrine, for all its faults, was held until the fourteenth century AD although, it is true, there were a few brave spirits who dared to question it. In the fourteenth century a theory of impetus developed, but its debt to the Indian theory is not clear. It *is* clear, however, that what the Indians proposed was a forerunner of what was later developed mathematically in the West during the Scientific Revolution. (Ronan 1983, pp. 194–195) [Ronan's emphasis]

Thus the impetus theory, which has been crucial to the rise of modern science, as many historians including Duhem, Butterfield, Westfall, and Kuhn have stressed, has many cultural roots. Although it is unclear whether the Indian theory influenced the West and, if it did, whether it was mediated through Arabic scholars, it is evident that Arabic studies of the impetus theory influenced European scientists. Moreover, it is possible that Arabic thinkers, discussing the problem of motion with Indian thinkers, would have learnt of the Indian theory—since they were well acquainted with Indian astronomical, mathematical, and medical lore. Ignoring the multicultural dimensions of the impetus theory would also lead us to align our historical narratives in a Eurocentric direction.

Another major influence on Copernicus, according to Kuhn, was the neo-Platonic vision of the cosmos. He argues that the neo-Platonic vision of the sun as the symbol of a fecund deity whose procreativity was demonstrated by the diversity and multiplicity of the forms in the cosmos, was a profound influence on Renaissance literature and the arts. Kuhn considers Copernicus to have been inspired by this neo-Platonic vision, and was led to conclude that it was only proper that the sun, given its significance as the source of light, warmth, and fertility should be at the center of the universe (Kuhn 1957, pp. 130–131). Copernicus himself informs us in the *Revolutionibus*:

> In the center of all rests the sun. For who would place this lamp of a very beautiful temple in another or better place than this wherefrom it can illuminate everything at the same time? As a matter of fact, not unhappily do some call it the lantern; others, the mind and still others, the pilot of the world. Trismegistus calls it a "visible god"; Sophocles' Electra, "that which gazes upon all things." And so the sun, as if resting on a kingly throne, governs the family of stars which wheel around. (Copernicus 1995, pp. 25–26)

However, it is strange that Kuhn should refer to Copernicus' view as neo-Platonic. It is more appropriately characterized as Hermetic. The Hermetic tradition, with the sun as the visible symbol of an invisible deity, developed out of Egyptian mystery religions in which the sun played an important role. Indeed shortly before Moses led the Jews out of Egypt the pharaoh Akhenaton had sought to impose a monotheistic doctrine with the Sun as the sole God of the universe throughout his empire. Although he failed in his religious revolution, the worship of the Sun as God, or visible symbol of the invisible God, was to resurface again and again in Egypt and elsewhere. In 274 CE in order to deal with the rise of Christianity the Roman Emperor Aurelian had proclaimed the Unconquered Sun (*Sol Invictus*) as the God of the Roman Empire.

In one of the earliest known Egyptian cosmologies—the Heliopolitan cosmology—the primeval God Atum, considered to be a manifestation of the Sun-Creator God, spat out the pair of gods Shu (air) and Tefnut (water) as the first act of cosmic genesis. Then Shu separated Nut (Heaven) from Qeb (Earth). From these five primal gods were born the other gods of Egypt—including Osiris, Horus, and Isis (Ronan 1983, p. 21). There is a double significance to this myth. In the first place the Sun God is made the progenitor god as in the Hermetic tradition. Moreover, the myth also expresses through personifications what were later to become the five elements of Platonic and Aristotelian cosmology—fire, earth, air, water, and *quintessence* are the Egyptian gods Atum, Qeb, Shu, Tefnut, and Nut, now seen impersonally as blind substances. It is possible—given the important influence of Egypt during the formative period of Greek thought, widely acknowledged by many Greek writers including Aristotle and Herodotus, and recently documented by Martin Bernal in his *Black Athena*—that the Heliopolitan cosmology is the original basis for the five-element theory of Greek science. This theory was to play an important role in Arabic and European science until the eighteenth century.

It is also known that the Hermetic tradition developed in Alexandria when Greek philosophy met Egyptian religion. Hence, it is probable that it was here that Egyptian religion and its cosmological doctrines, attributed to Hermes or Thoth, absorbed Greek philosophy. Similar assimilations were to occur in the future when Christianity and Islam were to find in Greek philosophy the language to formulate their theological visions. Augustine found in Plato the philosophical basis for Christian inspiration; for the Muslim *falsafah,* a Platonized Aristotelianism served the same purpose; and Aristotle became the major philosopher for the medieval scholastics. Just as it would be false to see Augustinian and Aquinian theology as neo-Platonism or neo-Aristotelianism, or the Islamic *falsafah* as merely offering a neo-Platonic Aristotelianism, thereby ignoring the religious vision that conditioned their thought, it is equally false to see Hermeticism as a variant of neo-Platonism, thereby erasing the fundamental Egyptian religious perspective that conditions it. The core vision of Hermeticism is the sun as visible symbol of an invisible deity—Platonism merely contributes the philosophical language for this vision.

Moreover, the Hermetic tradition traces its roots back to Hermes-Thoth, who is supposed to have lived around the time of Moses. This was also the period of the Akhenaton's attempted revolution, and it may reflect a conflict in Egyptian society itself as to whether there should be worship only of the one creator God—Atum—by whom all the other gods were created, or whether worship of the other gods should also be permitted. The monotheism of Akhenaton did not succeed but it is likely that the change itself influenced the Egyptian religious orthodoxy to place greater emphasis on the Sun as the primal god. When this new religion of the Sun absorbed Platonic ideas in Alexandria the basis for the Hermetic tradition was laid. If this argument is accepted then what Kuhn describes as the neo-Platonic influence on Copernicus should really be seen as the Egyptian influence on Copernicus—that is, the influence of Egyptian Sun-worship. This introduces a new multicultural influence on Copernicus—one that injects an African component to the Indian and Arabic influences we have already encountered.

In his study *The Egyptian Hermes*, Fowden lends credibility to the position adopted here. Until recently the general tendency has been to follow Yates and reject the view of Ficino and the Renaissance tradition that Hermetic ideas were rooted in Egyptian magical traditions, that went back to the time of Moses. Instead, it was assumed that they were really Hellenic philosophical texts formulated in late pagan Alexandria, during the first centuries of the current era. Although Fowden does not dispute the fact that they were written in the late pagan age in Egypt, he argues that they are really a synthesis of Greek and Egyptian traditions. He writes:

> When the two alien cultural traditions of Egypt and Greece began to mix, it was on terms that bore little relation to political realities. In the centers of power, Hellenism was triumphant; but in cultural terms Egyptianism, instead of being submerged by Hellenism, exercised so strong a gravitational and assimilative pull on it that the product of their interaction was at least as much Egyptian as Greek. Nowhere was this truer than in matters of religion. (Fowden 1986, p. 14)

Hence many major influences on Copernicus identified by Kuhn ultimately originated outside Europe—in Arabic physical realist critiques of both Ptolemaic theory and Aristotle's theory of projectile motion, and their attempts to develop alternative planetary and impetus theories; in Egyptian Hermetic influences associated with a solar religion; and in Indian mathematical techniques (even if Kuhn thinks that the mathematical techniques used by Copernicus only depended on Ptolemy). If we fail to take into account the wider dialogical context of these scientific ideas, and merely start at the point where these ideas are first found in Europe, we would create an illusory history of science that would see the new ideas as generated de novo within Europe. We would end up with a narrative similar to that given by Kuhn showing these ideas as originating in medieval scholastic critiques of Ptolemy, scholastic developments in impetus theory, and neo-Platonism.

Moreover, to take into account the role of the scholastics without the Arabic contribution is particularly misleading. The Arabic critiques of Ptolemy were more sustained, and developed with greater mathematical sophistication, than the scholastic ones. It is true that in the impetus theory, especially in relation to its application to explain the motion of the heavenly spheres, scholastic thought did develop in new directions with Buridan. But these can also be seen as building on work earlier done by the Arabic scientists.

Similarly describing Hermeticism as neo-Platonism also grossly distorts the picture, since the central ideas that shaped Copernicus were more rooted in Egyptian than in Platonic thought. Nowhere in Plato do we find the notion of the sun as a deity in the way it is found in Egyptian Hermeticism. As the quote above suggests it is belief in the preeminent status of the sun that inspires Copernicus to place it at the center by referring to it as "this lamp of a very beautiful temple," "the lantern," "the mind," and the "pilot of the world." He even appeals to the authority of Trismegistus, who had called it a "visible god," to justify its location at the center of the universe. The visionary language Copernicus uses, in the very text in which he presents his revolutionary theory, is closer to the spirit of ancient Egyptian Sun-worship than to Platonic rationalism. To describe Copernicus as motivated by a neo-Platonic influence, as Kuhn implies, is to give credit to the Greek philosophical language that cloaks the Egyptian religious vision, but to ignore the body of the vision altogether.

What is evident in Kuhn's historical approach is that in every one of the cases considered, it discusses and emphasizes the European context of the influence on Copernicus but ignores the wider multicultural context that shaped it. Thus Kuhn's narrative erases the dialogical standpoint it is possible to take on the history of the Copernican Revolution. Given the state of historical writing when Kuhn wrote, his Eurocentrism may be forgivable, but surely it is unacceptable today.

Kuhn also identifies another factor that promoted skepticism about Ptolemaic astronomy—the European voyages of discovery. These widened the geographical horizon of European thinkers—especially the discovery of America and the journey around the Cape of Africa—and showed how wrong Ptolemy's geography had been. If Ptolemy, considered the greatest geographer of antiquity, could have made gross mistakes in the geographical arena, it was possible to imagine that his views on astronomy could also be equally erroneous. Kuhn speculates that the voyages of geographical discovery may have made it easier for Copernicus, and other astronomers of the Renaissance, to question established astronomical ideas (Kuhn 1957, p. 124). However, we could add to Kuhn's speculations about the possible impact of the voyages of discovery. We have seen that these voyages exposed European scholars to new ideas from China, India, and other places quite unlike any they had confronted before. Could these ideas themselves not have created an ambient environment that made Europeans more receptive to thinking of alternatives to the Ptolemaic model? Needham emphasizes this possibility and suggests the following:

> In a word, specific investigations might be worth making to ascertain whether the complete disbelief of the Chinese in the solid celestial spheres of the Ptolemaic-Aristotelian world picture, which became evident to the Jesuits as soon as they began to discuss cosmology in China, was not one of the elements which combined in breaking up medieval views in Europe, and contributed to the birth of modern astronomy. (Needham 1958, pp. 5–6)

The broader implications of Needham's recommendation cannot be ignored when we consider these facts. When Ignatius Loyola, the founder of the Jesuit order, died in 1556, the Jesuits were operating a network of seventy-four colleges across the world—in Europe, Asia, and the Americas. Their *Collegio Romano* trained mathematically and scientifically proficient missionaries who were sent out to all parts of the world. It was headed by Christoph Clavius, the author of an important mathematical textbook for Jesuit education, who kept in regular contact not only with missionary scientists such as Matteo Ricci in China, but also with leading figures such as Galileo in Europe. Feingold notes:

> We are still far from fully conscious of the enormous contribution of Jesuit teachers to the formation of Catholic secular culture during the early modern period. That the Jesuit fathers cared for more than 200,000 children and adolescents each year is staggering in itself. But we may also recall that the Jesuits produced Torricelli, Descartes, Mersenne, Fontenelle, Laplace, Volta, Diderot, Helvetius, Condorcet, Turgot, Voltaire, Vico, and Muratori, to name but a few non-Jesuits. (Feingold 2003, p. 38)

Indeed if we take Needham's question seriously these facts should lead us to reexamine the Eurocentric assumption that the Jesuit transmission of knowledge was a one-way process from Europe to the rest of the world, whatever the self-perception of the Europeans may have been. It may also have been a dialogical process that brought knowledge from other parts of the world into Europe that crucially influenced the Scientific Revolution. Kuhn convincingly speculates about the possible implications of the European voyages of geographical discovery, but fails to see that they could also have been voyages of intellectual discovery that brought Europeans in contact with other cultural ideas.

Continuing to examine the theory proposed by Copernicus in detail Kuhn writes:

> Most of the essential elements by which we knew the Copernican Revolution—easy and accurate computations of planetary position, the abolition of epicycles and eccentrics, the dissolution of the spheres, the sun a star, the infinite expansion of the universe—these and many others are not to be found anywhere in Copernicus' work. In every respect except the earth's motion the *De Revolutionibus* seems more closely akin to the works of ancient and medieval astronomers and cosmologists than to the writings of the succeeding generations who based their work upon Copernicus' and who made explicit the radical consequences that even its author had not seen in his work. (Kuhn 1957, p. 134)

Kuhn concludes that the Copernican theory itself was not revolutionary even though it had revolutionary consequences. These changes came about only as a result of making explicit the radical implications inherent in the theory—these constituted the real intellectual revolution and were the unanticipated by-products of the Copernican transposition of the center of the universe from the earth to the sun. The small change made by Copernicus in astronomical theory forced others to rethink Aristotle's explanation of why planets moved the way they did in the heavens, and why bodies near the earth exhibited the sort of local motions that they were seen to have. This led to changes in our understanding of celestial and terrestrial dynamics, and constituted the wider Copernican Revolution that ended with Newton. Hence, Kuhn argues, we have to consider *De Revolutionibus* more as a revolution-making than as a revolutionary work (Kuhn 1957, p. 135). He then proceeds to trace in detail the development of this wider Copernican Revolution that led to the birth of modern science.

He begins with Tycho Brahe. Kuhn emphasizes that the significance of Brahe lies in the extensive, accurate, and up-to-date astronomical data he collected and not in any theoretical contribution he made. He was responsible for many of the important changes made in observational techniques, and the more rigorous standards of precision demanded of astronomers in the modern era. One important innovation introduced by him is the habit of making observations regularly along the whole length of the path of a planet in the sky, rather than only at locations convenient for observation. He was thus able to furnish a vastly expanded and dependable body of evidence for future theoretical astronomers. The significance of his contribution to the science of observational astronomy can be appreciated when we consider that Kepler could not have discovered his laws of planetary motion if he had depended on the more inaccurate corpus of data available prior to Brahe's studies.

Nevertheless, what is lacking in Kuhn's account is an appreciation of the important contributions made by Arabic astronomical instruments and techniques to Brahe's achievements. Many of these were developed within the tradition of Arabic astronomy and it is significant that even the observatory that Brahe built was modeled on Arabic precedents. According to Nasr:

> In fact it can be said without exaggeration that the observatory as a scientific institution owes its birth to Islamic civilization. While in the early Islamic period the observatory was of a small size and usually associated with a single astronomer, from the 7th/13th century and the building of the Maragha observatory by Nasir al-Din al-Tusi, it became a major scientific institution in which numerous scientists gathered to work and teach together. (Nasr 1976, p. 20)

The Maragha observatory became the model for the Ulegh-Beg observatory in Samarqand and the Istanbul observatory in Ottoman Turkey. These in turn became the models for Brahe's observatory which, both in architectural style and the instruments deployed, was influenced by its Arabic precedents.

There are remarkable similarities between the instruments described by al-Urdi, and by Taqi al-Din at the Istanbul observatory, and those described by Brahe in his *Astronomiae Instauratae Mechanica*. The mural quadrant that Tycho Brahe prided in and called *Tichonicus* was also built earlier by Taqi al-Din, who referred to it as the *libnah*. Nasr concludes that "a close study of later Islamic instruments reveals the astonishing degree to which early European observatories followed Islamic models" (Nasr 1976, p. 126).

At the same time we cannot separate Brahe's emphasis on precise and careful observation from the revolution in art and anatomy in the Renaissance period. We have seen that perspectival representational painting, and Vesalius' accurate depiction of the human body, were both inspired by the transformation in European consciousness largely triggered by Alhazen's optical revolution. Hence if we follow Kuhn's narrative and ignore the Arabic roots of the theoretical and instrumental basis for Brahe's achievements we would distort our historical narrative in a Eurocentric direction.

We have already seen that there is the possibility of a Chinese influence on Brahe when he became the first astronomer in Europe to adopt an equatorial mounting for his telescope. In his case, there was no theoretical basis for adopting this mounting although such a mounting would have been natural for Chinese astronomers, who considered the heavenly bodies to rotate around the pole star—they would, in effect, be taking into account the actual rotation of the earth, which is the basis of the illusory rotation observed around the pole star. It would also make sense to adopt such a mounting if we subscribe to the Copernican theory, since it assumes the earth to rotate. However, Brahe believed in neither the Copernican theory nor the Chinese theory. He did, however, construct an alternative geoheliocentric model of the universe in which the inferior planets—Mercury and Venus—revolved around the sun, but the sun and the other superior planets revolved around the earth. However, his theory took the earth to be stationary. Hence Brahe had no theoretical motivation for being either led to discover, or adopt, the equatorial mounting for his telescope.

This makes it possible to suspect that Brahe could not have been led to select the mounting as a result of being guided by how he considered the heavens to move. Unless we propose that he discovered it accidentally, we have to assume that he came to know about the Chinese style of mounting astronomical observational instruments and discovered, by trial, that it provided a more convenient way of making regular observations. Even his practice of making systematic observations of planetary motions on a daily basis followed the Chinese pattern. Working more than half a century after the Portuguese arrived in China it is possible that European astronomers may have become well aware of the rigorous and systematic manner in which the Chinese collected astronomical data. Indeed only Chinese civilization, apart from the Mesopotamian and Arabic, kept accurate and detailed astronomical records before modern times—and in the precision and faithfulness with which they recorded the diverse events in the heavens the Chinese remained unmatched.

In 1572 Brahe observed a nova in the sky that dimmed and vanished after eighteen months. He showed through accurate observation that this was a superlunary phenomenon—which put into question the Aristotelian cosmological assumption that no change took place above the lunar sphere. Brahe also observed comets—in 1577, 1580, 1585, 1590, 1593, and 1596. He showed that comets were superlunary events since their paths cut through the crystalline spheres. Kuhn argues that Brahe's discovery, that comets were not terrestrial exhalations but celestial events, played an important role in promoting the transition to the Copernican worldview (Kuhn 1957, pp. 206–208).

However, what Kuhn ignores—in fact does not even address—is the fact that Chinese astronomers had long held the view that comets were celestial events, and, as we have seen, kept faithful records of hundreds of comets over a long period of time. Could the Chinese view of comets have influenced Brahe to see them as really extraterrestrial phenomena? We have to remember that Brahe's observations were made nearly sixty years after Portuguese contact with China in 1514. Otherwise we have a number of strange coincidences—Brahe's observational practices in astronomy involving systematic daily recording of astronomical data, his technique for mounting telescopes, his receptivity to change above the lunar sphere (all remarkably close to Chinese observational practices and beliefs, and involving significant breaks with the European tradition) that we have to attribute, as Kuhn does, to the "genius" of Brahe. To this list we could add Brahe's geo-heliocentric theory, which paralleled the Nilakantha planetary theory formulated in 1501. Surely it is more reasonable to suppose that Brahe could not have been uninfluenced by the corridor of communication established between Europe and Asia at the end of the fifteenth century.

Brahe's astronomical data were used by Kepler to derive the three laws of planetary motion that came to play a crucial role in Newton's final consolidation of the wider Copernican Revolution. What led Kepler to give up the geocentric model and turn to the Copernican theory? Moreover, what led him to believe that the planets orbiting the sun were obedient to simple mathematical laws? Kuhn traces Kepler's inspiration to neo-Platonism. According to Kuhn "the mathematical magic and the sun worship that are so marked in Kepler's research" (Kuhn 1957, p. 132) shows contact between Renaissance Hermeticism and the new astronomy conditioned by a Platonic mathematical interpretation of Hermetic symbolism.

Kepler was an ardent Neoplatonist. He believed that mathematical simple laws are the basis of all natural phenomena and that the sun is the physical cause of all celestial motions. Both his most lasting and his most evanescent contributions to astronomy display these two aspects of his frequently mystical Neoplatonic faith .... It also played an immensely important role in his own research, particularly in his derivation of his Second Law upon which the First depends. In its origin the Second Law is independent of any but the crudest sort of observation. It arises rather from Kepler's physical intuition that the planets are pushed around their orbits by rays of a moving force, the *anima motrix*, which emanates from the sun. These rays must, Kepler believed, be restricted to the plane of the ecliptic, in or near which all the planets moved. Therefore the number of rays that impinged on a planet and the corresponding force that drove the planet around the sun would decrease as the distance between the planet and the sun increased. At twice the distance from the sun half as many rays of the *anima motrix* would fall on a planet, and the velocity of the planet in its orbit would, in consequence, be half of its orbital velocity at its original distance from the sun. (Kuhn 1957, p. 214)

It was from this early speed-law that Kepler derived his Second Law—namely, that the line joining the planets to the sun in their elliptical motion sweeps out equal areas in equal times. Though the Second Law is not equivalent to the speed-law since the latter is only approximately true, the speed-law served to guide Kepler toward the correct law of areas. In this sense, according to Kuhn, neo-Platonic inspiration played a crucial role in Kepler's derivation of his first two, and ultimately third, laws of planetary motion.

However, a closer examination shows that neo-Platonism cannot bear the weight Kuhn attempts to place upon it. If by neo-Platonism it is suggested that the root ideas are traceable back to Plato then this is certainly false. The centrality of the sun is not a part of Plato's cosmology—rather it is, as we have seen, more closely linked to Egyptian cosmology mediated through Hermeticism. Indeed a more comprehensive reading of Kepler's views shows it to be more in accord with Hermetic Sun-worship than the Platonic search for forms. Consider the following passage from Kepler quoted by Kuhn:

> [The Sun] is a fountain of light, rich in fruitful heat, most fair, limpid and pure to the sight, the source of vision, portrayer of all colors, though himself empty of color, called King of the planets for his motion, heart of the world for his power, its eye for his beauty, and which alone we should judge worthy of the Most High God, should he be pleased with a material domicile and choose a place in which to dwell with the blessed angels … [Hence] by the highest right we return to the sun, who alone appears by virtue of his dignity and power, suited for this motive duty and worthy to become the home of God himself not to say the first mover. (Kuhn 1957, p. 131)

This passage puts Kepler, like Copernicus as we saw earlier, closer to the ancient Egyptian adoration of the sun than to the views of Plato. Moreover, if by neo-Platonism is meant the desire to find a mathematically precise and physically true account of the motion of the planets then this ideal cannot be traced back to Plato either. Indeed it is in Ptolemy, who was satisfied with "saving the phenomena" without regard to physical plausibility, that we find the faithful expression of Plato's view that perfect mathematical forms cannot be realized in this world. Hence, Kepler's mathematical realism is really rooted in the Arabic orientation to astronomy—it was the Arabic scientists who pursued the goal of a mathematically precise and physically true model of the world. Moreover, in Alhazen's optical theory they offered an exemplar that met these conditions.

Kepler's notion that rays from the sun influence the planets cannot be traced back to Plato either. Moreover, it cannot be traced to any Arabic or Hermetic ideas. In Hermeticism the sun may be considered to be a source of influence on the planets, but this influence was not seen as rays radiating out from the sun in the form of lines. This idea appears to be derived from Chinese notions of lines of influence connecting different planetary bodies like the magnetic lines of influence that guided the orientation of the lodestone on earth. According to Needham there was indeed such an influence on Kepler:

> Magnetic science was indeed an essential component of modern science. All the preparation for Peter of Maricourt, the greatest medieval student of the compass, and hence the ideas of Galileo and Kepler on the cosmic role of magnetism, had been Chinese. Gilbert thought that all heavenly motions were due to the magnetic powers of heavenly bodies, and Kepler had the idea that gravitation must be something like magnetic attraction. The tendency of bodies to fall to the ground was explained by the idea that the earth was like an enormous magnet drawing things unto itself. The conception of a parallelism between gravity and magnetism was a vitally important part of the preparation for Isaac Newton. In the Newtonian synthesis gravitation was axiomatic, one might almost say, and spread throughout all space just as magnetic force would act across space with no obvious intermediation. Thus the ancient Chinese ideas were a very important part of the preparation for Newton through Gilbert and Kepler. (Needham 1969, pp. 73–74)

If Needham is right then the reason why Kepler treats the influence from the sun as radiating out toward the planets is because he conceived them in terms of the Chinese idea that it is lines of connections between heavenly bodies which allow them to move in harmony with each other.

Hence what Kuhn describes misleadingly as the neo-Platonic influence on Kepler is really compounded of separate influences from three different cultures—the Egyptian religious influence mediated through Hermeticism, which assigns the sun a central role in the universe and views it as having a controlling effect on the planets; the Chinese idea of lines of connections between heavenly bodies that determines the specific way in which the sun influences the planets; and the mathematical realist orientation of Arabic thinkers, which assumes that actual planetary motions can be described in precise mathematical language and not merely in a language that "saves the phenomena." By making neo-Platonism carry the load of these triple multicultural influences Kuhn renders a Eurocentric picture of the dialogical birth of modern science.

According to Kuhn the victory of the Copernican theory in the long run was assured once Kepler showed that each planet moved in a simple ellipse and obeyed elegant mathematical laws. No such simplification of astronomical theory would have been possible in the Ptolemaic model of an Earth-centered universe. The complex system of eccentrics, epicycles, and equants used by Ptolemy for each planet was no longer needed. Moreover, Kepler's accomplishment also made it possible for him to predict the motions of the planets far more precisely than hitherto possible. This would also have ensured the victory of the Copernican theory over time. Hence, Kuhn concludes, we arrive with Kepler at the point where the Copernican theory would ultimately have been accepted by leading astronomers even if there was no new evidence in favor of it. Nevertheless, new evidence did come from an unexpected direction—the telescope (Kuhn 1957, p. 219).

In 1609 Galileo turned the telescope upon the heavens. It was an instrument he had constructed himself based on reports that a Dutch lens grinder had been able to magnify distant objects by combining two lenses. In the process he opened up a new cosmos never before seen. First, in every direction he looked Galileo found new stars—he discovered hundreds of thousands over and above the few thousand stars known before. The idea of an infinite universe no longer seemed implausible. He also discovered that although the number of stars seen increased dramatically with the telescope, the size of the stars did not—they continued to appear as points of light. This could only be explained if they were much farther away than astronomers had hitherto suspected. It further resolved one major objection to the Copernican theory—the absence of stellar parallax despite the enormous distances traversed by the earth as it orbited the sun.

Galileo also discovered that the moon was not a crystalline sphere but covered with pits and craters, valleys and mountains. It was not perfectly spherical as Aristotelians maintained heavenly bodies should be. The telescope revealed the sun to have spots (sunspots) that moved in a fashion that could only be explained if the sun were rotating. The rotation of the sun lent some credibility to the notion that the earth could also rotate. Even more significant was the discovery of the four principal moons of Jupiter that orbited the planet—this provided visual evidence that heavenly bodies could orbit other heavenly bodies (and not only the earth, as the Ptolemaic theory assumed). Hence nothing precluded the possibility that the planets revolved around the sun, as Copernicus maintained.

However, Kuhn rightly argues that none of the data provided evidence in direct support of the Copernican theory—they could have been accommodated into the existing Ptolemaic model. The model could have been saved by giving up the idea that the heavens were unchanging and by adding the new stars to those already known in the stellar sphere. Even the moons of Jupiter favored neither Ptolemy nor Copernicus because the issue between them was whether the earth or sun was to be the center of the universe—and the appearance of a new center in Jupiter for heavenly orbits was neutral to this issue (or could be taken as subverting both) (Kuhn 1957, p. 224).

There was, however, one observation that favored the Copernican theory—the planet Venus, seen through the telescope, exhibited phases like the moon. On the Ptolemaic model the appearance of Venus should always be in the form of a crescent since the fully lit face of the planet could not be seen at any point of its orbit from Earth. By contrast, the Copernican theory predicted that at times Venus and the earth would be opposite each other relative to the sun, at which time Venus' whole lit surface would face the earth.

Copernicus himself had noted that the appearance of Venus relative to the sun could provide evidence about its motion. However, his argument is concerned more with showing that this view could reveal whether the orbit of the planet is below or above the sun, but is not offered as evidence for his theory—either alternative is compatible with the Ptolemaic theory. In his *Revolutionibus* he writes:

> In addition, there is the fact that they [Mercury and Venus] are small bodies in comparison with the sun, since Venus even though larger than Mercury can cover scarcely one one-hundredth part of the sun, as al-Battani the Harranite maintains, who holds that the diameter of the sun is ten times greater, and therefore it would not be easy to see such a little speck in the midst of such beaming light. Averroes, however, in his paraphrase of Ptolemy records having seen something blackish, when he observed the conjunction of the sun and Mercury which he had computed. And so they judge that these two planets move below the solar circle. (Copernicus 1995, pp. 20–21)

However, the telescope showed that Venus sometimes came between the earth and sun, and that sometimes the sun came between the earth and Venus—showing that Venus orbits the sun. This refutes the judgments of al-Battani and Averroes that Mercury and Venus always moved below the solar circle.

Nevertheless, Kuhn warns us against overestimating the evidence provided by the telescope. He writes:

> The evidence for Copernicanism provided by Galileo's telescope is forceful, but it is also strange. None of the observations discussed above, except perhaps the last [i.e., the phases of Venus], provides direct evidence for the main tenets of Copernicus' theory—the central position of the sun or the motion of the planets about it. Either the Ptolemaic or the Tychonic universe contains enough space for the newly discovered stars; either can be modified to allow for imperfections in the heavens and for satellites attached to celestial bodies; the Tychonic system, at least, provides as good an explanation as the

Copernican for the observed phases of and distance to Venus. Therefore, the telescope did not prove the validity of Copernicus' conceptual scheme. But it did provide an immensely effective weapon for the battle. It was not proof, but it was propaganda ... the Copernicans, or at least the cosmologically more radical ones, had anticipated the sort of universe that the telescope was disclosing. They had predicted a detail, the phases of Venus, with precision. More important, they had anticipated, at least vaguely, the imperfections and the vastly increased population of the heavens. Their vision of the universe showed marked parallels to the universe that the telescope made manifest. (Kuhn 1957, p. 224)

Significantly, Kuhn's account makes no mention of the theory that made the telescope possible—Alhazen's optical theory. This theory had not only influenced earlier medieval thinkers such as Witelo, Roger Bacon, and Peckham, who essentially adopted it as their paradigm of optics, but had also continued its influence to include Galileo and Kepler (Nasr 1976, p. 140). Indeed Galileo used it to determine the heights of mountains and depths of craters on the moon revealed by the telescope. This raises the question: What effect did Alhazen's theory have on the development of the telescope?

It is noteworthy that a new translation of Alhazen's *Optical Thesaurus* appeared in Basle in 1572. In the work he discusses phenomena such as the behavior of light reflected off parabolic and spherical mirrors; the principle of "least time" for the path of light rays (often associated with the name of Fermat); the application of the rectangle of velocities at the surface of refraction (associated with Newton); and numerous studies of refraction in glass cylinders immersed in water. More significantly, for our purpose, he also discusses magnifying lenses in this work (Nasr 1976, p. 141). Thus Alhazen not only furnished the theoretical basis for the behavior of light in the telescope, but also discussed the magnifying effect of the lenses that came to be used in it.

Taking this into account, and the fact that it remained the dominant optical paradigm until the time of Kepler, should we not consider Alhazen's theory to have played an important role in the design of the telescope, and thereby, on the whole history of the discoveries in the heavens it made possible? Indeed it is conceivable that without the theory and Alhazen's study of the behavior of magnifying lenses, the emergence of the modern telescope may have been unlikely. If this is the case then the theoretical discoveries of Arabic science can be considered to have laid the basis for some of the most important instrumental innovations in modern science, such as the telescope and microscope—both of which opened up new dimensions of the universe with far-reaching consequences for the future history of science.

Finally Kuhn turns to the culmination of the wider Copernican Revolution in the Newtonian synthesis. He argues that two different historical paths led from the achievements of Copernicus, Kepler, and Galileo to Newton. The first is the atomic or corpuscular philosophy; the second is the conception of force able to answer the question, What moves the planets? (Kuhn 1957, p. 243). He considers that the corpuscular view was the first to consolidate as a well-defined approach to physical phenomena, although the concept of force—albeit not in the form it ultimately took in Newton's theory—can be discovered in Kepler's concept of *anima motrix*. Kuhn writes:

> Early in the seventeenth century atomism experienced an immense revival. Partly because of its significant congruence with Copernicanism and partly because it was the only developed cosmology available to replace the increasingly discredited scholastic world view, atomism was firmly merged with Copernicanism as a fundamental tenet of the "new philosophy" which directed the scientific imagination. Donne's lament that because of the "new philosophy" the universe was "crumbled out again to his Atomies" is an early symptom of the confluence of these previously distinct intellectual currents. By 1630 most of the men who dominated research in the physical sciences showed the merger's effects. They believed that the earth was a moving planet, and they attacked the problems presented by the Copernican conception with a set of "corpuscular" premises derived from ancient atomism. (Kuhn 1957, p. 237)

Descartes was an important figure in leading this new approach to solving the problems raised by the Copernican theory, that fused atomism with the mechanical philosophy. He extended the idea of mechanism to the microlevel in order to make the two approaches compatible. Beginning by asking how a single free corpuscle would move in space, Descartes proceeded to consider how its motion would then be affected by collisions with other particles. To the first question he provided a successful answer—a free particle would move in a straight line without any change of velocity. According to Kuhn the Cartesian answer was "a straightforward consequence of the impetus theory." However, Kuhn seems to have underestimated the significance of the change introduced by Descartes. In Buridan's impetus theory the motive power imparted to the body decreased as it continued its motion. With Descartes no motive power was required to keep a body in motion. This was an important change because it meant that motion was a natural state of a body requiring no explanation; only alterations of motion require explanation. Hence what Descartes proposed was really an inertial account of motion—one that is distinct from an impetus account.

Descartes then proceeded to develop his idea further by deriving laws that constrain the particles in collisions. However, in this regard he failed, for only one of the seven laws he proposed came to be accepted by later scientists. Moreover, in order to explain the motion of large bodies in space Descartes made his space full by packing his particles so tightly together that there was no empty space between any of them. In effect he created a particulate fluid *plenum*.

The notion of a fluid medium in Cartesian space seems to have Arabic precedents—we have seen that Thabit Ibn Qurra constructed a planetary model in which all the spaces between celestial spheres were filled with a fluid. However, there is a profound difference between the Cartesian and Ibn Qurra approaches. By beginning with a void that was gradually filled with particles until they were packed so tightly that they squeezed the void out altogether, Descartes was able to deploy an atomic mode of analysis to his fluid *plenum*. Although the Cartesian *plenum* was ultimately seen as initiating an unsustainable research program, Descartes' approach to problems in physics and cosmology in terms of particles in collision—that is, his mechanical atomic orientation—was profoundly influential.

Where did Descartes get the atomic idea? Kuhn assumes that the atomic views of the seventeenth century were inspired by ancient Greek and Hellenistic theories. By contrast we have suggested that the modern atomic view could have been influenced by Arabic *kalam* and, possibly, Indian atomism. Hence, it is intriguing to find the historian of science Alexandre Koyré also expressing doubt about the notion that modern atomism is a revival of the ancient Greek theory:

> The atomism of the ancients, at least in the aspect presented to us by Epicurus and Lucretius—it may be that it was different with Democritus, but we know very little about Democritus—was not a scientific theory, and though some of its precepts, as for instance, that which enjoins us to explain the celestial phenomena on the pattern of terrestrial ones, seem to lead to the unification of the world achieved by modern science, it has never been able to yield a foundation for development of a physics; not even in modern times: indeed, its revival by Gassendi remained perfectly sterile. The explanation of this sterility lies, in my opinion, in the extreme sensualism of the Epicurean tradition; it is only when this sensualism was rejected by the founders of modern science and replaced by a mathematical approach to nature that atomism—in the works of Galileo, R. Boyle, Newton, etc.— became a scientifically valid conception, and Lucretius and Epicurus appeared as forerunners of modern science. It is possible, of course, and even probable, that, in linking mathematism with atomism, modern science revived the deepest intuitions and intentions of Democritus. (Koyré 1957, p. 278, n. 7)

Thus Koyré suggests that the success of atomic doctrines in the seventeenth century and their sterility in the Hellenic world can be attributed to the role of mathematics in modern atomism. However, he does not explain why the mathematical methods and techniques were not deployed in ancient times—he merely takes it to be an unfortunate contingent fact of history. But the Greek and Hellenistic scientists could not have deployed mathematics with their atomic theories even if they had wanted to. It was the Indian number system, and the algebra linked to it, that made the application of mathematics to atomic theories possible. Moreover, there are strong reasons to suspect that the affinity of the new system of mathematics to atomic views was not merely a fortuitous accident—the number system and algebra had both been originally developed in close association with atomic ideas in Indian culture.

To see this point let us begin with what Descartes has to tell us about the influence of algebra on the development of his ideas. In his *Discoveries on the Method of Rightly Conducting the Reason and Seeking for Truth in the Sciences,* Descartes contrasts the abstract and rigid approach of the geometric methods of the ancients with the more concrete and flexible, but complex, approach of the algebraic methods available in his time:

> And as to the Analysis of the ancients and the Algebra of the moderns, besides the fact that they embrace only matters the most abstract, such as appear to have no actual use, the former is always so restricted to the consideration of symbols that it cannot exercise the Understanding without greatly fatiguing the Imagination; and in the latter one is so subjected to certain rules and formulas that the result is the construction of an art which is confused and obscure, and which embarrasses the mind, instead of a science which contributes to its cultivation.[4]

He then proceeds to show how the situation can be remedied by borrowing the best in geometrical analysis and algebra and combining them so that the defects of one become corrected by the other.[5] This synthesis was, of course, the Cartesian discovery of coordinate geometry, in which geometrical straight lines, circles, ellipses, and other figures could be represented by algebraic functions. The path Descartes followed can be seen as the inverse of that taken by Omar Khayyam and al-Kashi, who sought geometrical forms to represent algebraic equations. In contrast to their geometrization of algebra, Descartes algebraized geometry. This interpretation of his approach shows the limits of the orthodox perception that Descartes modeled his discoveries only on Euclidean geometry. This becomes more evident when we look at his account of the methodological revolution he made:

> It is the same as with a child, for instance, who has been instructed in Arithmetic and has made an addition according to the rule prescribed; he

may be sure of having found as regards the sum of figures given to him all that the human mind can know. For, in conclusion, the Method which teaches us to follow the true order and enumerate exactly every term in the matter under investigation contains everything which gives certainty to the rules of Arithmetic.

But what pleased me most in this method was that I was certain by its means of exercising my reason in all things, if not perfectly, at least as well as was in my power. And besides this, I felt in making use of it that my mind gradually accustomed itself to conceive of its objects more accurately and distinctly; and not having restricted this Method to any particular matter, *I promised myself to apply it as usefully to the difficulties of other sciences as I had done to those of Algebra.*[6] [My emphasis]

It is significant that Descartes should refer to algebra as inspiring the development of his method as well as his insights into geometry. There is indeed a close link between algebra and arithmetic in the Indian decimal place number system with zero. Every number can be represented in the form of a power series—for example, $3725 = (3 \times 1000) + (7 \times 100) + (2 \times 10) + 5 = 5 + 2.10 + 7.10^2 + 3.10^3$. This is easily extended to series such as $a_0 + a_1 x + a_2 x^2 + a_3 x^3 + \cdots$—a generalization Indian mathematicians did go on to make to arrive at the concept of a power series. Just as a number can be represented as a power series based on the number 10, so is a power series a representation of a number on a base that could be any number x. This can be extended to represent trigonometric functions as power series—precisely the route taken, as we saw earlier, by Indian mathematical astronomers of the Kerala School. The basic principle adopted in every such representation is that any number or function can be represented as the sum of parts. For rational numbers the sum involves a finite series; for irrational numbers an infinite series. Nevertheless, the idea that the whole can be calculated as a sum of parts is built into the arithmetical number system, and the algebraic system discovered as its generalization, by Indian mathematicians. Significantly the same idea plays an important role in integral calculus, where the area within a figure is treated as the sum of infinitesimal parts—an idea the Indian mathematicians developed, although they did not reach the modern concept of a limit function.

The notion that a whole can be treated as the sum of its parts is crucial for the development of seventeenth-century physics. Many of the fundamental concepts and laws of physics discovered at that time embody this additive principle—the mass of a body is the sum of the masses of each of its parts; the momentum of a body is the sum of the momenta of each of its parts; the volume, force, distance moved, et cetera, of any body is the sum of the volumes of its parts, the different forces on it, or the separate small intervals it traverses. Thus the mathematical system inherited from the Indians provided a powerful instrument for the atomic mode of analysis central to the development of seventeenth-century science. This atomism did not apply to physical bodies alone—it also applied to properties of these bodies such as volume, mass, length, momentum, force, and so on. Even time could be seen as the sum of separate temporal intervals. If we had to find a term to characterize this flexible atomic notion then the expression "mathematical atomism" could serve to distinguish it from the narrower physical or material atomism associated with Hellenic thought. Indian number theory embodied mathematical atomism and served as the mathematical foundation to study a vast range of properties of bodies and their behavior. Even integral calculus could be seen as the outcome and natural extension of the mathematical atomism embodied in the Indian number system since it computed areas, volumes, and so on by dividing them into tiny infinitesimal parts that are then summed together.

How did the number system of the Indians come to possess this mathematical compatibility with corpuscularism or material atomism? One important clue is that the representation of numbers within the Indian tradition developed in close conjunction with the pervasive Indian atomic orientation to nature. In both Hellenic and ancient Chinese culture the notion of a world made of atoms was never a dominant idea. In India the atomic idea goes back to the time of the Buddha, and came to be adopted at a very early stage by Hindu, Buddhist, and Jain thinkers. Even where they differed with regard to the nature of the fundamental atoms out of which the world was made there was consensus on the atomic orientation in general—so much so that atomism was built into the metaphysical basis of the religious worldviews of Hindus, Buddhists, and Jains in the way the *yin-yang* principle, or the notions of the five-element theory, became a part of Chinese and medieval scholastic beliefs.

For example, the Hindu Nyaya Vaisesika consider the four elements—earth, air, fire, and water—to be made of indestructible and indivisible atoms. These atoms are seen to be spherical in shape and in constant motion. They can combine into dyads, and such dyads can in turn combine to form triads—each part of the triad being a dyad. Deploying such concepts—reminiscent of the ways modern atoms combine to form molecules, and these in turn larger molecules—the Nyaya, albeit with a more restricted notion of the ways atoms can combine, were able to explain many of the physical and chemical properties of things in the world (Goonatilake 1984, p. 17). It is evident that in the manner they used the concept of atoms the approach of the Nyaya is more sophisticated than that of the Greek atomists. This is not surprising because atomism was a more ancient idea in India, constituted a dominant tradition there, and underwent extended development over long periods of time.

In contrast to the Hindu view that there were different kinds of atoms associated with different elements, the Jains considered all atoms to be identical. They explained the different properties of the elements by arguing that these resulted from different combinations of atoms. Moreover, the Jains also attributed properties such as attraction and repulsion to atoms. The sophistication of their approach becomes evident when we notice that they not only considered atoms in objects to sometimes vibrate, but also maintained that free atoms traveled in straight lines.

In conformity with their metaphysics of process and incessant change, Buddhist thinkers pictured atoms as bundles of forces or energy. They maintained that atoms had no permanent existence but appeared and disappeared continually to be replaced by other similar atoms. Their view reminds us of *kalam* in the Arabic world—it is possible that *kalam* atomism itself, as we saw earlier, could have been influenced by such Buddhist notions. However, the appearances and disappearances of the Buddhist instantaneous point atoms does not occur in isolation from other atomic occurrences as *kalam* holds—instead it is held to follow the Buddhist law of dependent origination so that each appearance is conditioned by the context in which it appears, and is a rigidly regulated causal process. Like the Jains, the Buddhists also consider atoms to be centers of repulsive and attractive forces for other atoms. For example, earth atoms repel other atoms; water atoms attract them. (Goonatilake 1984, p. 17)

We have seen that Koyré attributed the success of modern atomism to the application of mathematics in its formulation. Hence it is significant that the Indian number system, which furnished many of the algebraic tools that made this possible, developed in close association with Indian atomism. It enabled Indian scientists to work with indefinitely small numbers when they dealt with the dimensions of atoms, and with indefinitely large numbers when they came to computing the number of atoms in large bodies. The mathematical atomism built into the number system made it an ideal instrument for the atomistic metaphysics that informed all the major Indian traditions of thought, and provided a way to represent the very small and very large numbers required by them. This came to be a precious gift to the seventeenth-century scientists in Europe who developed modern science: the mathematical atomism that informed Indian studies in the number system, infinite series, and proto-calculus meshed in extremely well with the general mathematical realist epistemology and mechanical-corpuscular ontology that European scientists set out to articulate.

Thus it is reasonable to suppose that atomism became a dominant idea in the seventeenth century because its potential could be fully articulated only with the new mathematical instruments of arithmetic, algebra, and trigonometry that European thinkers inherited from their Indian predecessors.

When Descartes absorbed Greek geometry into algebra he combined the power of the Greek deductive geometric approach with the mathematical atomism implicit in the Indian mathematical instruments. He united the axiomatic rationalism of the Hellenic tradition with the mathematical atomism of the Indian tradition. If atomism in the seventeenth century appears to have suddenly become an idea that caught on it was largely because it was not only carried surreptitiously in the mathematical instruments of the new number system, and the algebra and trigonometry that developed out of it, but it also connected historically to the substance, spatial, and temporal atomism of Indian thinkers who inspired it.

This is not to deny that the ideas of *kalam* and the atomism of al-Razi played a role in influencing the atomic turn. However, without the mathematical atomism implicit in the new Indian mathematics it is not likely that atomism could have taken off in the modern era in the way it did. Moreover, even in the case of *kalam* and al-Razi, as we have seen, the Indian influence cannot be excluded. This suggests that Indian atomism as ontology might have entered Europe through *kalam,* and Indian atomism as a mode of mathematical analysis through Indian mathematics.

The final consolidation of the wider Copernican Revolution was achieved by Isaac Newton. It was he who drew together the corpuscular ideas inherent in Descartes' particulate *plenum,* the laws of planetary motion discovered by Kepler, and Galileo's laws of terrestrial motion through his novel concept of gravitational force—a force that drew the planets to the sun and the moon to the earth, and caused bodies to fall to the earth in local motion. Newton's concept of gravitation was controversial because it violated a fundamental principle of the new mechanical philosophy that there cannot be action-at-a-distance unmediated by material contact—an occult notion associated with the despised tradition of astrology, and instantly criticized by leading thinkers such as Leibniz.

Newton himself repudiated the charge that gravitation was an occult notion. He argued that he was merely concerned with a mathematical account of motion and claimed to "feign no hypotheses" about the nature of gravity. However, Kuhn argues, Newton was dissatisfied with his answer because it did not give a complete explanation of the phenomena observed. He writes:

> Descartes's corpuscles had been totally neutral; weight itself had been explained as a result of impact; the conception of a built-in attractive principle operating at a distance therefore seemed a regression to the mystic "sympathies" and "potencies" for which medieval science had been so ridiculed. Newton himself entirely agreed. He repeatedly attempted to discover a mechanical explanation of the attraction, and though forced at last to admit his failure, he continued to maintain that someone else would succeed ...

It does not, I think, misrepresent Newton's intentions as a scientist to maintain that he wished to write a *Principles of Philosophy,* like Descartes, but that his inability to explain gravity forced him to restrict his subject to the *Mathematical Principles of Natural Philosophy.* Both the similarity and the difference of titles are significant. Newton seems to have considered his magnum opus, the *Principia,* incomplete. It contained only a mathematical description of gravity. Unlike Descartes's *Principles* it did not even pretend to explain why the universe runs as it does. (Kuhn 1957, p. 259)

However, what Kuhn overlooks in this account is that Newton himself adopted the position that gravity needed an explanation only when he was criticized by the mechanical philosophers. It is more likely that Newton believed in the reality of the so-called occult force of gravity since, as we found earlier, he believed that God in the beginning had created particles "moved by certain active principles, such as is that of gravity, and that which causes [chemical] fermentation, and the cohesion of bodies" (Newton 1952, p. 401).

Newton's belief in such forces may have been shaped by his early interest in Hermetic ideas. Newton was interested in alchemy all his life; and the alchemical tradition in Europe was strongly influenced by Hermeticism. For the alchemists gold was the purest metal, which they symbolized by the sun. Alchemists held that the processes they studied involved the action of active principles that united and separated the elements in chemical processes. The historian of science Morris Berman maintains that Newton was a closet Hermeticist and that the mechanical philosophy adopted by Newton was merely the garb in which he presented himself in public. Berman writes:

What Newton did, then, was to delve deeply into the Hermetic wisdom for his answers, while clothing them in the idiom of the mechanical philosophy. The centerpiece of the Newtonian system, gravitational attraction, was in fact the Hermetic principle of sympathetic forces, which Newton saw as a creative principle, a source of divine energy in the universe. (Berman 1984, p. 115)

Newton's hidden sympathy for Hermetic ideas becomes more evident when we consider that he is known to have on occasion traced the roots of his scientific views to Egypt—the birthplace of Hermeticism. The point here is not whether Newton was right—it is highly unlikely that the ancient Egyptians ever entertained the idea of universal gravitation—but rather that Newton's belief that Egyptian priests knew the secrets of the cosmos revealed in the *Principia* is an indication of the deep influence of Hermetic ideas upon him (Berman 1984, p. 116). Hence Newton is likely to have been attracted by the alchemist view that elements are driven to combine by sympathetic forces, and that the sun, which symbolizes gold, should have the special status that Copernicus assigned it. It explains how the notion of gravitation and that of a sun-centered universe may have become linked in Newton's mind.

It may be supposed that the notion of gravitational force can be traced back into the Chinese idea of lines of influence. This view is not tenable because Newton's force is not mediated over time but acts directly with one body influencing another instantly. The Chinese idea is of a field-like influence spreading out over time. It was dissatisfaction with the Newtonian concept of action-at-a-distance that led Einstein to a field conception of gravity more than 200 years later.

However, most historians of science have often been reluctant to acknowledge the profound impact of Hermetic ideas on Newton's thinking. In his study *From Paracelsus to Newton: Magic and the Making of Modern Science* Charles Webster notes:

> The magnitude of evidence indicative of the tenacity of interest in philosophies running contrary to the mechanical philosophy is so great that the only way of accommodating this vast anomaly has been to separate the leaders of science—judged representative men of their age—from the unrepresentative and more gullible majority. It is unfortunate for any proponent of this line that figures of outstanding importance, including Newton himself, turn out to display a lively interest in the occult. The only means of saving the phenomenon in this case is to adopt the unconvincing device of postulating a split personality for the scientists convicted of lapsing from consistent practice of the enlightenment ideal. (Webster 1982, p. 3)

Nevertheless, Webster notes that the economist Maynard Keynes, after studying the extent of Newton's alchemical interests, and the vast bulk of his writings in this area, was forced to conclude that Newton was not the first of the modern rationalists, but the last of the magicians (Webster 1982, p. 9).[7]

There is also another more insidious consequence of treating Newton as a split personality or of seeing him simply as the last of the magicians who incidentally, in an act of supreme forgetfulness, also created modern science. It masks the fact that the Newtonian achievement was influenced crucially by thematic ideas of Egyptian religion carried by the Hermetic tradition. The fact that so many of those who made seminal contributions to the Scientific Revolution, such as Copernicus, Kepler, and Newton, were so profoundly affected by Hermetic ideas should give us pause. Indeed it is possible to conclude that by ignoring the role of Egyptian Hermetic influences on Copernicus' adoption of the heliocentric theory, Kepler's support for it, and Newton's concept of force that linked it to the physics of motion, and tracing all these influences to Kuhn's catch-all notion of "neo-Platonic influence," we would again historically erase another profound dialogical impact on modern science—an impact not from Asia but African Egypt.[8]

If the arguments presented above are accepted then the Copernican Revolution—both in the narrow sense of the heliocentric revolution of Copernicus, and in the wider sense of the Scientific Revolution that culminated in Newton—has to be understood as a complex process involving influences from a plurality of cultural traditions. The Arabic scientists and thinkers paved the way by their mathematical realist critiques of the Ptolemaic model; by the exemplar of a mathematical realist theory they provided with Alhazen optics and its influence on Renaissance science and aesthetics; and by their critiques of Aristotelian conceptions of causality, culminating in the notion that the regularities of nature are "the habits of God." They made it possible for the conception of empirically discoverable, but not rationally deducible, mathematical laws of nature to arise in the modern era.

From the Chinese came a stream of mechanical technologies whose swift transfer shifted European sensibilities so that they came to perceive the world itself in the image of a machine. Moreover, the Copernican theory made it possible for Chinese cosmological ideas to enter modern astronomy and open the door to a new universe of infinite empty space populated by stars, comets, and novae in incessant transformation to replace the unchanging finite *plenum* of Aristotle containing planets and stars embedded in rotating crystalline spheres.

From India modern science received powerful mathematical ideas and techniques—a new flexible number system, advanced algebraic methods, trigonometry, and possibly the beginnings of calculus. The mathematical system itself carried the new mode of atomic thinking that reconditioned the way nature came to be studied. Moreover, implicitly through the mathematical instruments transmitted, and possibly also more directly through the influence of *kalam* and al-Razi, Indian atomism came to nurture modern science with one of its most seminal notions.

Finally from Egypt came Hermetic ideas (fructified by Platonic mathematical idealism and philosophic notions) of the central status of the sun in the heavens and that changes in nature were wrought by active principles. These notions inspired many of the pioneering figures of the Scientific Revolution—Copernicus, Kepler, and Newton. In Newton the notion of a sun-centered universe and gravity as an active principle united to consolidate modern science.

However, all of these streams came together only as a result of the Copernican theory. The Copernican vision acted as the attractor that drew ideas from many different traditions to resolve the problems it raised concerning terrestrial and celestial motions. In a sense Kuhn is right to argue that the Scientific Revolution was largely the unanticipated by-product of the Copernican theory. However, it is unlikely that the revolution would have taken the direction it did, or be so swiftly consolidated, were it not for the multicultural resources upon which European scientists and scholars could draw.

Nevertheless, the revolution itself was not a mere adding of notions from diverse cultures. The makers of the revolution—Copernicus, Kepler, Galileo, Descartes, Newton, and many others—had to selectively appropriate relevant ideas, transform them, and create new auxiliary concepts in order to complete their task. To say that they drew on other traditions outside their culture is no more to impugn their creative achievement than to say they drew on a reservoir of Hellenic and Hellenistic ideas. Neither does it diminish the European synthesis in Europe of the new science—any more than Arabic or Indian achievements are diminished because they drew on Greek ideas. In the ultimate analysis, even if the revolution was rooted upon a multicultural base it is the accomplishment of Europeans in Europe. The dispute is not about who made the Scientific Revolution or where it happened; the dispute is about whether modern science drew upon European or multicultural ideas. We have argued that global traditions of science came together at the birth of modern science.

Chapter 14

Contrasting Competitive Plausibility

Throughout this study we have seen that themes from many different cultures played a significant role in the genesis of modern science. Nevertheless, the history of modern science continues to be conditioned by a Eurocentric perspective in which non-European cultures are perceived to have made little or no contributions, and many of their influences are, instead, traced back to minor themes in the ancient Greek tradition. An illusory history of science is thereby constructed, which continues to shape thinking even today. Thus the biologist Lewis Wolpert writes:

> The peculiar nature of science is responsible for the fact that, unlike technology or religion, science originated only once in history, in Greece. Most scholars are agreed that science had its origin in Greece, though those that equate science with technology would argue differently. This unique origin is important for understanding the nature of science, since it makes science quite different from so many other human activities, for no other society independently developed a scientific mode of thought, and all later developments in science can be traced back to the Greeks. (Wolpert 1993, p. 35)

Such views are by no means confined to scientists—even Thomas Kuhn, in his work *The Structure of Scientific Revolutions,* which can be seen as one of the most important epistemological influences on the recent multicultural turn in many areas of knowledge, appears to become equally Eurocentric when he considers the history of science:

> Every civilization of which we have records has possessed a technology, an art, a religion, a political system, laws and so on. In many cases those facets of civilizations have been as developed as our own. But only the civilizations that descend from Hellenic Greece have possessed more than the most rudimentary

science. The bulk of scientific knowledge is a product of Europe in the last four centuries. No other place and time has supported the very special communities from which scientific productivity comes. (Kuhn 1970, pp. 167–168)

As revealed in the passages above the Eurocentric intellectual history of science is, to a large extent, really a history of the Greek roots of modern science. Other histories, equally Eurocentric, which trace the roots of science to Europe in the scholastic age, are not so much intellectual as cultural histories, see the medieval period as formative for the political, social, or economic conditions that facilitated the rise of modern science. Duhem is one exception—he traced the rise of modern science to Buridan's impetus theory and Oresme's geometric interpretation of certain arithmetical values, which he considers to have paved the way for the law of inertia and Cartesian analytic geometry in the seventeenth century. However, most historians of science have been reluctant to follow Duhem and locate the birth of modern science in the medieval fourteenth century. Indeed the discoveries of Buridan and Oresme only continue the work done by Arabic studies of the impetus theory and geometric approaches to algebraic problems. If Duhem is right, then we would also have to locate the birth of modern science in the Arabic world, but this would be to ignore the distinctive features of modern science that make it different from the tradition of Arabic science.

The main reason Eurocentric interpretations appear so plausible is that many of the major themes that influenced modern science can indeed be traced back to Hellenistic thought—we find the application of mathematics to natural phenomena in Ptolemy and Archimedes, the interest in mechanical systems in Archimedes, the Hermetic tradition in Alexandrian Hellenistic thought, the atomic theory in Leucippus and Democritus, and so on. However, explaining modern science as a process of continuing the developments initiated in Hellenistic science creates a number of problems. First, this tradition was also inherited by the Arabic and Byzantine cultures. The Hellenistic tradition was carried by them for several centuries (in the case of the Byzantines, for more than a millennium). We need to develop auxiliary constructions to explain why the tradition remained so sterile in these two cultures—insofar as the emergence of modern science is concerned—but took off suddenly in modern Europe once it was transmitted there in the scholastic period.

Second many of the themes of modern science can also be traced back to Chinese and Indian thought. Mathematics played an important role in Chinese astronomy (especially algebra) and was a highly developed science among Indian mathematical astronomers. The Chinese had developed far more powerful mechanical technologies than the Greeks. The sophistication of Indian atomic theories was only exceeded in the nineteenth century, and, even in China, atomic theories were proposed by the Mohists although they did not develop far. Hence one could, in principle, trace most of the relevant themes found in Hellenistic thought to both the Indian and Chinese traditions simply by virtue of the prolific range of ideas generated there in the Axial Age.[1] Hence only the fact that modern science developed in Europe (which happened to include Greece) makes it seem reasonable to trace current scientific ideas to the Axial Age Greek world. If modern science had developed in China or India, then their Axial Age thinkers would have provided sufficient material for Sinocentric or Indocentric histories analogous to the Eurocentric account that dominates thought today.

This suggests that Greek science did not really play the unique role now assigned by historians to it. It came to be seen as the only predecessor of modern science simply because it possessed a sufficiently rich reservoir of ideas for historians to construct it as such by selective appropriation of themes guided by the hindsight knowledge provided by modern science. This process is analogous to the way theological hermeneutics finds the discoveries of modern science in selected sacred texts— the texts treated as significant are now Greek philosophical and scientific texts. The suspicion that this may be the case becomes reinforced when we consider that Greek science did not develop into modern science in all of the cultures that inherited it before Western Europe—the Hellenistic world itself, as well as the Byzantine and Arabic civilizations. This triple failure of the tradition to generate modern science can, of course, be explained by appeal to sociocultural or religious factors formulated in the form of answers to the question, Why did modern science not develop in civilization X? However, we have already had occasion to show the questionable status of the question. It is far easier to suppose that Greek culture was never on the verge of modern science than to find sociocultural or religious obstacles in diverse cultures for their perceived failure to develop modern science in spite of having at hand the resources to do so.

Hence, the mere presence of themes found in modern science in a particular culture—like the Greek—cannot be sufficient to claim that they inspired a similar emergence in modern science without other conditions being met. These themes must be developed sufficiently to become useful to the creators of modern science; and they have to be selected and fused together into a coherent new structure. This study has argued that Arabic thinkers, rather than the Greeks, made mathematical realism a dominant theme, and provided an exemplar of a mathematical realist theory in Alhazen optics. It was this achievement that made mathematical realism seem a plausible approach to nature. Second, it was the Chinese who developed mechanical technologies to a level where a mechanistic conception of the universe could become plausible, and it was Indian mathematical atomism (inspired by Indian atomism) that made possible the atomistic analysis of matter, its properties, space, and time at the dawn of modern science. Moreover, the peculiar emphasis on sun-centeredness and vital forces in Egyptian Hermeticism motivated the urge to see the sun as the center of the planetary system, and inspired Newton in his conception of a gravitational force that drew diverse bodies in the universe to one another. The unique European achievement in the modern age—an age formed by the achievement itself— was to formulate the heliocentric theory, draw on highly developed themes from a diversity of cultures, and integrate them into a new science able to consolidate the theory.

Hence, the European achievement is more radical than Eurocentric interpretations suggest—it did not merely continue the Greek heritage of science, or for that matter the Arabic, Chinese, Indian, or Egyptian traditions, but broke with all of them by creating a science forged from elements drawn from every one of them. This would not require us to deny the contributions of Hellenic and Hellenistic science; it would only put into question Eurocentric histories that one-sidedly stress the Greek roots of modern science, but ignore other cultural contributions. Such histories underestimate the uniqueness and novelty of modern science by reducing it to an outcome of Greek science. Our historical account is more faithful to the perception of the creators of modern science that they were engaged in a war of the ancients and moderns in which ancient Aristotelian science, assimilated only a few centuries earlier, had to be subverted to make way for modern science.[2]

If the birth of modern science can be attributed to the confluence of four streams of science—the Arabic, Chinese, Indian, and Egyptian— entering Europe from outside, then it can also explain why modern science developed within Europe and not elsewhere. There is no need to appeal to sociocultural values in non-European cultures that led to their failure to arrive at modern science. Many of the debates on this question have focused on Chinese and Arabic science because they are considered to have been sufficiently advanced to have produced modern science, but did not. The blame for the Chinese failure has been attributed to various cultural, social, and institutional causes, but there is a more direct explanation. The Chinese never acquired the Arabic tradition of science as the Europeans did, and therefore they were in no position to combine a mathematical realist orientation with their mechanical achievements to create modern science.

It has also been argued that Arabic science was on the threshold of modern science simply because it had inherited Greek science. However, while Arabic civilization did inherit the Greek corpus, as well as the Indian and the Hermetic traditions, it did not absorb Chinese science, and especially its technology, to the same extent as Europe. After the Mongol destruction of the Abbasid Caliphate the Islamic world collapsed into disarray (exacerbated by the terrible plague that followed), and perpetual invasions by Mongols, Turks, and other nomadic people created too much instability for Chinese technology to be absorbed into productive activity—even though the invaders themselves converted to the religion of Islam. Hence, the Renaissance combination of artist and craftsman, like Leonardo da Vinci, who united the mathematical realist approach of Arabic science with interest in mechanical devices that were ultimately of Chinese-inspired origin, could not develop here. Such artist-craftsmen laid the basis for the mathematical mechanical vision of the future.

The failure of modern science to develop in India can be explained by the straightforward fact that it was not an arena for the influx of Chinese technology. Instead India in the north came under Islamic rule in the eleventh century, and the turmoil of the Arabic world after the Mongol conquests did not facilitate the flow of Chinese technologies from the north. Hence one important factor that favored Europe over the Arabic world and India was the impact of Chinese technology there—without which the mechanical vision of the universe would not have emerged.

The multicultural impact on Europe would also explain why Greek science failed to develop into modern science. It could not have—it lacked Indian mathematics, and it lacked the mathematical realist orientation of the Arabic scientists, as well as the significant exemplar of a mathematical realist theory of science provided by Alhazen's optics, and Chinese mechanical technologies. Given this vast abyss separating Greek science from modern science, it simply did not have the resources to develop into modern science. Moreover these resources were not even available at the time Greek science began to decline—most of the advances in the Arabic, Indian, and Chinese traditions that made modern science possible came centuries after the decline of Hellenic science.

Consider Cohen's argument that intellectual factors alone cannot have led to the breakthrough to modern science in Europe because these factors were also present in the Arabic world. To explain why the advance happened in Europe and not in the Arabic world we have seen Cohen appeal to what he describes as the "European Coloring." He lists the factors that went into this "Coloring" as follows: the urge to make very accurate observations of natural phenomena; the tradition of the application of mathematics to art by Renaissance artists; the positive valuation of manual labor in Europe; a greater receptiveness to magical notions and practices; and the receptivity to corpuscular views, which fortunately integrated extremely well with the universe of mathematical precision.

However, the dialogical approach this study has adopted allows us to recognize that these factors did not fortuitously arise in Europe de novo, as Cohen supposes. They were the outcome of Europe's interaction with other cultural influences. The urge for accurate observation and application of mathematics to art was largely inspired by the impact of the Arabic world (and particularly the Alhazen optical theory); the positive valuation of manual labor came from the recognition of the productive possibilities inherent in the mechanical technologies from China; the receptiveness to magical notions and practices is traceable to the impact of Egyptian Hermetic philosophy; and the union of the corpuscular philosophy with the universe of precision was made possible by Indian mathematical theories. Although it could be argued that the "European Coloring" was different from the Islamic one around 1600, the difference itself can be explained not by Eurocentric appeal to some mysterious essence present in Europe from ancient times, but as the outcome of changes wrought in Europe through its dialogue with Asian and African cultures.

However, we cannot ignore the role of the Copernican theory in functioning as the attractor around which the different multicultural streams could coalesce. The mathematical realism of Arabic thinkers was drawn into it by the insistence that astronomical theory should provide both a mathematically precise and physically plausible account of planetary motion. Chinese astronomical ideas became a part of this theory as it expanded to include evolving stars, comets, nova, and other parts of a changing universe of objects in a space construed as both infinite and empty. Indian mathematical ideas were drawn into it through the work of Indian astronomers, and these ideas also facilitated the ontological and analytical atomism of modern science. And Egyptian Hermetic conceptions of the sun as deity inspired its central heliocentric idea and the concept of gravitational force. The Copernican theory was a distinctively European achievement—linked to old Europe through Aristarchus and new Europe through Copernicus. Given its power to focus relevant elements of other traditions into its research program, it also explains why modern science developed in Europe and not elsewhere.

Thus this dialogical account is able to give one explanation for why modern science developed in Europe and not elsewhere. First, it was in Europe that the Copernican theory developed. Hence in no other culture did such an attractor for multicultural traditions exist. Second, it was in Europe that it became possible for the interaction of Arabic, Chinese, Indian, and Egyptian sciences to occur. By contrast Eurocentric accounts of modern science require one set of factors to explain why it developed in Europe, and another set to explain why it failed to emerge elsewhere—with a separate set of specific factors to account for the failure of Arabic, Chinese, and Indian sciences, respectively. Hence, in terms of economy of thought and explanation the dialogical model is more competitively plausible than the Eurocentric one.[3]

However, the dialogical answer to Needham's Grand Question also has much wider cultural implications. Today we find ourselves in the middle of a so-called "science wars," with feminists, environmentalists, and multiculturalists turning to premodern traditions to rectify perceived limits in the system of knowledge generated by modern science. To those who subscribe to the Eurocentric vision of the history of science, such attempts might appear perverse movements inspired by antiscience sentiments returning to myth, superstition, and irrationality. However, if modern science did progress through dialogue with premodern traditions of knowledge—indeed its birth itself was the outcome of such a process—then there is good reason to suppose that dialogue with these traditions may tap yet unexploited reservoirs of knowledge for future science. Hence the dialogical history of science opens the door to dialogical approaches to science in the future. A Eurocentric history does not give us any reason to open the door to a dialogue; worse, it may lead us to shut it. Hence the science wars are not just about the history of science—they are also about the future of science.

Notes

CHAPTER 1

1. Such critics are normally inspired by the postmodern philosophies of science associated with Kuhn (1970) and Feyerabend (1975). Sandra Harding (1998) uses these epistemological reorientations to argue for a more multicultural approach to the history and philosophy of science. See also Ross (1996) for a collection of articles concerned with defending feminist, multicultural, and environmental perspectives on current science.

2. The call for greater acknowledgment of multicultural traditions is often inspired by environmental, health, and sometimes feminist concerns. Nandy (1988) brings together a number of writers who wish to construct an alternative science and technology that would promote wider recognition for traditional systems of knowledge that they view as struggling against the hegemony of modern science. Caroline Merchant (1980) calls for a return to more holistic premodern traditions of knowledge to deal with environmental concerns. Morris Berman (1984) also thinks that such a return can resolve the psychological wounds suffered by humans as a result of their alienation from nature that modernity imposes. Shiva (1988) combines multicultural, feminist, and environmental critiques of science together in an attempt to expose the limits of modern science.

3. The literature of this genre has also burgeoned. See, for example, the scathingly critical views of postmodern and multicultural thought in Gross and Levitt (1994) and Gross, Levitt, and Lewis (1996). Weinberg (2001) and Wolpert (1993) are two leading scientists who have also addressed such concerns in some depth. A broader critical assessment of postmodern critiques of science can be found in Koertge (2000), Holton (1993), Kurtz and Madigan (1994), and Nanda (2003). Nanda also views with consternation the alliance of postmodern and premodern movements as a result of their shared opposition to modernist views associated with science.

4. More recently there have been attempts to open science epistemologically to a dialogue with other traditions and to go beyond the conflicts inherent in the science wars. For such approaches see Brown (2001), Ashman and Baringer (2001), and Carrier et al. (2004).

5. The dialogical perspective on the history of science can be said to have begun systematically with Needham's series on *Science and Civilization in China* (Needham 1954, 1956), which showed the numerous historical contributions

of Chinese science to modern science. Similar attempts to document the Arabic and Indian contributions were made by Nasr (1968) and Bose, Sen, and Subbarayappa (1971), though not on the same scale. Diop (1991) worked in isolation over many decades to make the case for an Egyptian–African impact on Greek science and, indirectly, on modern science. Joseph (2000) is probably the first attempt to systematically document the dialogue of multiple traditions of science, in this case mathematics, as they historically interacted with one another and also paved the way for modern science.

6. The term Arabic is used to describe the civilization that developed under the impact of Islam but that also included many non-Arab ethnic communities (Persians, Turks, Europeans, and Indians among others), and many non-Muslim religious communities (including Christians, Jews, and Zoroastrians). It stresses the fact that much of the scientific work of the civilization created as a result of the spread of Islam was written in Arabic, and even those who wrote in other languages had their studies translated into Arabic for wider dissemination. To appreciate some of the problems associated with labeling civilizations in general, and Arabic/Islamic civilization in particular, see Joseph (2000), pp. 349–352.

7. See Wolf (1982), Abu-Lughod (1989), Amin (1989), Blaut (1993), Goody (1996), Frank (1998), Pomeranz (2000), and Hobson (2004) for studies of the global sociocultural context of the rise of modern technology and society. These studies go beyond the narrow Eurocentric orientation of most sociological studies of the growth of modern knowledge and institutions by looking at the global influences that shaped European thought in the modern era.

8. In this context it is interesting to note that although postmodern thought promotes greater receptivity to the historical idea that non-Western cultures made significant contributions to modern science, it also promotes resistance to the notion that different cultures confront the same world. Most postmodern views treat multicultural perspectives as incommensurable with one another and, therefore, render them unable to develop critiques of alternative frameworks or symbiotically integrate such frameworks with each other. What they do offer is a critique of hegemonic claims made by any framework—including that of modernity. This makes it difficult to develop any critique of modernity from a postmodern perspective that would subvert its specific beliefs—except to make modernity one possible belief system among others. If this is the case, then postmodernists implicitly provide, as Vattimo argues, a defense of "weak modernity"—i.e., modernist views without hegemonic claims. See Vattimo (1988).

9. For a more detailed study of how Western technology drew on the resources of non-European cultures even in the modern era, see Dharampal (1971), Alvares (1979), Pacy (1990), and Hobson (2004).

10. Nevertheless, the alliance of postmodernists and traditionalists is often fraught with tension. Although both sides are united in their opposition to the hegemonic claims of modern science, the pluralist relativism of postmodern theory does not sit easy with traditionalist foundational beliefs. Traditionalist sympathies can be found in Alvares (1979) and Sardar (1988)—the former arguing for paying more regard to traditional tribal perspectives, and the latter to religious perspectives. Nanda (2003) highlights the tension between postmodern

relativism and traditionalist absolutism despite their shared opposition to the hegemonic claims of modernity.

11. The significance of a dialogical perspective is now being increasingly recognized as important for advancing science, peace, and sustainable growth. The first year of the twenty-first century was declared by the United Nations to be the *International Year of Dialogue Among Civilizations.* Moreover, the International Council of Scientific Unions (ICSU) and the International Union for the History and Philosophy of Science (IUHPS) have both developed position papers that support dialogue with traditional cultures to advance scientific knowledge. It is particularly in the areas of environmental knowledge, medicine, agriculture, and peace that dialogue is seen as most important. Thus historical studies that show how scientific knowledge grew in the past through a dialogue of civilizations provide support for those who wish to encourage such dialogue in the future. For more on UN efforts to develop dialogical perspectives, see UNESCO's *Dialogue Among Civilizations* site: http://www.unesco.org/dialogue2001/.

12. Surprisingly, despite his involvement with issues in biotechnology, Fukuyama does not address the issue of traditional knowledge directly—nor does he examine recent literature concerned with tapping traditional environmental knowledge to advance science. His blindness in this regard is linked to his modernist belief that current science, and the method it has found, will drive the directionality of future history, and that other traditions of local knowledge are destined to fade away as influential factors in history.

13. Huntington does not think that the adoption and application of science requires Western values, and he does not see the possibility of a dialogical negotiation of shared values for all civilizations. Hence his moral perspective may be seen as indirectly consonant with a postmodern perspective, since it assumes incommensurable moral systems across civilizational divides. He splits the world into eight civilizations—Chinese, Japanese, Indian, Islamic, Western, Orthodox Russian, Latin American, and African—but sees the history of the West as essentially disconnected from the Rest though, paradoxically, its science is seen as acceptable to the Rest. Hence it is their religious-normative differences that lead him to see the future in terms of an emerging clash of civilizations.

14. Lynne-White first used the expression "Needham's Grand Question," albeit in passing. However, Needham picked up on it and considered it "one of the greatest problems in the history of civilization" (Needham 1969, p. 148). More recently, the historian of science Floris Cohen has identified it as crucial for understanding the birth and growth of modern science (Cohen 1994, p. 381, n. 5).

15. The importance, for civilizational relations, of the answer to Needham's Grand Question is evident once we see that it has an intimate connection with Weber's Grand Question, "Why did capitalism/modernity develop in the West and not elsewhere?" A dialogical answer to Needham's question would suggest that civilizations can mutually enrich one another and that intercivilizational relations need not be exclusivist or adversarial. See Nelson (1974) for a more detailed study of the connection between Needham's and Weber's central

questions concerning civilizations. It is interesting that Weber's modernist standpoint leads him to answer his question in terms of what keeps civilizations apart; Needham's dialogical standpoint shows how civilizations have in the past enriched each other—suggesting that they can do so in the future too.

CHAPTER 2

1. There have been many attempts to answer this negative version of Needham's Grand Question, especially in relation to the Chinese, Indian, and Arabic traditions. Answers have been as diverse and varied as attempts to answer Weber's question concerning why modern capitalism emerged in the West and not elsewhere. For attempts to develop an answer for the Chinese case see Needham (1969), Graham (1989), Bodde (1991), Sivin (1984); for the Arabic case see Saunders (1963) and Hoodbhoy (1991); for the Indian see Sangwan (1991). Also Cohen (1994) (especially in Chapter 6) raises it in the broader context of both Chinese and Arabic cultures and gives a primarily internalist answer; Huff (2003) does the same for the two cultures but gives an externalist explanation. Both essentially offer Eurocentric responses.

2. This does not mean that the question is never asked. Nevertheless, whatever answer is given is moderated by the perception that Greek science is a predecessor and precondition for modern science. Farrington (1949), e.g., argues that Greek science underwent spectacular development as a community of seafarers and merchants, free from the constraints of mythical thought and political authoritarianism, turned to the study of the natural world, but the subsequent development of the institution of slavery made it decline as theoretical thinking became sundered from practical concerns. See also Bernal (1971).

3. The same point is repeated by the biologist Lewis Wolpert, who writes:

> The peculiar nature of science is responsible for science having arisen only once. Even though most, if not all, of Aristotle's science was wrong—he can even be thought of as the scientist of common sense—he established the basis of a system for explaining the world based on postulates and logical deduction. This was brilliantly exploited by Euclid and Archimedes. By contrast the Chinese, often thought of as scientists, were expert engineers but made negligible contribution to science. (Wolpert 1993, p. xii)

Intriguingly when Einstein was asked why science did not develop in China, he avoided answering the question by proclaiming that the wonder was that it developed in Europe. He appears to have understood it to be a one-off miraculous event that could not be explained (Gillespie 1960, p. 9).

4. See Blaut (1993) for a critique of such views. He argues that Eurocentric history sees scientific knowledge as something generated within Europe and diffused elsewhere, and explains Europe's creativity in terms of racial, religious, cultural, and environmental factors. Although appeal to the first two factors has now become unfashionable, culture and environment are still considered

important. In a later study, Blaut (2000) examines the Eurocentrism of eight leading historical thinkers, including Max Weber, Lynn White, Jared Diamond, and David Landes.

5. According to Hobson (2004), such Eurocentric approaches to understanding how ideas, technology, and institutions have historically progressed is endemic to all the dominant orientations in the social sciences, including Marxism/systems theory, liberalism, and Weberianism. All assume that "Europe autonomously developed through an iron logic of immanence." He sees the most significant change in current historical studies as the shift away from debates engaging Marxism/world systems theory, liberalism, and Weberianism to "Eurocentrism vs anti-Eurocentrism" (p. 3).

6. Ronan's study (1983) is particularly useful for our purpose because it summarizes much of the literature that developed over the three decades following the pioneering studies of Needham on multicultural traditions of science. Other studies since then, focussed not on the content of scientific beliefs but advances in technology. See e.g., Pacy (1990) and McClellan and Dorn (1998). Selin's edited volume on *Science, Medicine and Technology* addresses the content of theories, but is not written from one integrated perspective. Unique in this regard is George Joseph's *Crest of the Peacock: Non-European Roots of Mathematics* (2000), which gives an integrated perspective on the content of multicultural mathematics.

7. Bodde's answers can be seen as the direct outcome of attempting to answer the Grand Question by beginning with the modernist assumption that the Chinese only had a pseudoscience, and rejecting Needham's conclusion that they had a protoscience that had contributed to modern science. Although Bodde begins his study by saying that his sole mission is to determine what factors in Chinese civilization facilitated or obstructed scientific and technological progress (Bodde 1991, p. 5), the rest of his study proceeds without reference to any actual Chinese science. Rather it is focused on the dynamics of written Chinese; Chinese ways of ordering space, time, and things; government and society; morals and values; and relations with nature, with emphasis on how these Chinese orientations served to obstruct the growth of science, but not technology.

8. Although the "structure of attitude and reference" that Said refers to is examined within the context of the history of the Western novel, and the ways in which it depicts non-European cultures, his approach also illumines the perceptions and value assigned to natural knowledge (i.e., scientific knowledge) accumulated by these cultures. See especially pp. 89–91. Said's views can be extended to criticize attempts to develop philosophies of science that treat traditional knowledge as pseudoscience. Such views are particularly insidious if they permit exploitation of traditional knowledge without proper acknowledgment. See Shiva (1997).

9. The point has also been made by Sivin. In attempting to explain the persistence of the question, Why did the Scientific Revolution not take place in China? he writes:

This way of putting the problem contains and supports certain Western assumptions, assumptions that ordinarily we do not question. Above all we usually assume that the Scientific Revolution is what everybody ought to have had … There is usually the further assumption that civilizations which had the potential for a scientific revolution ought to have had the kind that transpired in the West and encompassed the sorts of institutional and social changes that are identified with modernization. These assumptions are usually linked to a belief—or a faith, if you prefer—that European civilization all along was somehow in touch with reality in a way no other civilization could be, and that its great share of the world's wealth and power came from some intrinsic fitness to inherit the earth that was there all along. (Sivin 1984, p. 537)

10. It is also noteworthy that the distinctively different evaluations of Needham, Ronan, and Bodde arise from the different judgments they make concerning the *theoretical* contributions of Chinese science to modern science. Needham argues that these contributions were significant in astronomy, magnetic science, mathematics, medicine, and other areas. Both Ronan and Bodde see Chinese contributions as primarily technological. The theoretical views of the Chinese, based on the *yin-yang* and five element theories, are seen as historically irrelevant to modern science by Ronan even though they informed traditional Chinese science and, as pseudoscientific notions, irrelevant to the growth of technology, even within traditional China, by Bodde. It is only Needham's dialogical view that modern science grew by learning from Chinese science that leads him to value it in the way Greek science has traditionally been. For Needham Chinese science, like Greek science, is a stage in the development that led to modern science.

11. Baber (1996) argues that the science and technology tradition in India came to be dramatically altered under colonial rule. Moreover, even if Indian science contributed to modern science, it nevertheless came to be largely displaced by it. Dharampal (1971) showed, by studying British documentary archives, that a flourishing Indian tradition of science and technology had been carefully studied by British colonial scientists to advance modern science and technology. However, he also argued that the same tradition was later repressed by colonial education, so that its dialogical contribution was rendered invisible.

12. George Sarton (1975, p. 17) had dated the glorious years of Arabic science between 750 CE and 1100 CE. Sabra (1988) sees the achievements as beginning in the middle of the eighth century and continuing to the fifteenth. Saunders (1963) argues that Arabic science had grown until the Mongol invasions, and declined after the collapse of the Abbasid Caliphate with the destruction of Baghdad. These different views can be reconciled if we divide Arabic science into two phases—an earlier phase inspired by Greek science and a later phase with distinctively new developments that broke away from Hellenistic science. The later phase—associated with Alhazen optics, the Maragha School of astronomy, al-Nafis' physiological studies—became intensely critical of Greek science. It was also during this period that we see the rise of *kalam* philosophy and its vehement critique of the Greek-inspired *falsafah* tradition.

13. Good exemplars of this receptivity include the studies of al-Andalusi (1029–1070) in the West and al-Biruni (973–1048) in the East in the eleventh century. What is distinctive about both these Arabic thinkers are their historical and epistemological perspectives—they saw Arabic science as having grown through the dialogical interaction of different cultures. Such a dialogical understanding of the birth and growth of modern science is only a recent development that began with Needham, but has yet to displace dominant Eurocentric explanations involving appeals to *immanent* intellectual and socio-cultural factors within Europe. See Alberuni (1983), al-Andalusi (1991).

14. Merton (1970) developed a suggestion by Weber, of a possible connection between the Protestant ethic and modern science, to argue that Puritan values were closely associated with the rise of modern science. His approach connecting religious and cultural values to answer the question of why science was born in Europe has also been used to explain why it did not arise elsewhere. See Huff (2003, pp. 22–25) for a discussion of Merton's core values—universalism, communalism, disinterestedness, and organized skepticism—that he takes to have made possible the rise of modern science in the West.

15. A similar approach has been made by Huff (2003) to explain why modern science did not develop in the Arabic or Chinese worlds. He traces the core factors that facilitated the rise of science as rooted in the encounter of Greek philosophy, Roman law, and Christian theology so that "together they laid the foundations validating the existence of neutral institutional spaces within which intellects could pursue their guiding lights and ask the most probing questions." By contrast, such freedom of thought was precluded from arising in the Chinese and Arabic worlds (pp. 362–363). Huff's account exemplifies how the positive and negative versions of Needham's Grand Question can be linked together to show how core values within cultures promoted and inhibited the rise of modern science. However, it is not only Western writers who are tempted to adopt such perspectives, as modernizing writers from Arabic, Indian, and Chinese cultures have also argued along similar lines. See Hoodbhoy (1991), especially Chapters 9–11, on the long battle for scientific rationality against religious orthodoxy within Islam throughout history; Qian (1985) on the factors that promoted scientific stagnation in traditional China; and Nanda (2003) on "prophets facing backwards" who obstruct science by returning to traditional knowledge.

16. However, it is important to note that it is not the ideal of a universal science that is at the root of the problem, but the misdirected assumption that modern science is universal science. Indeed the universal ideal of science can coexist with the notion that many local traditions of science have contributed, and can in future contribute, to universal science. Universal science would then develop through a dialogue between many local traditions of science, each helping to define the scope and limits of others. However, if modern science is treated as universal *per se*, then it becomes impossible to take critical perspectives on it from the context of other traditions of knowledge, and it only leaves scope for one-sided critical perspectives of other traditions from the vantage of modern science.

17. In an oft-quoted passage in the Analects, when asked about serving spirits, Confucius said, "If we are not yet able to serve man how can we serve spiritual beings." [Quoted in (Chan 1969, p. 36)] Contrast this attitude with the sentence imposed on Socrates by Athens for disbelieving in the gods or believing in false gods. Hence, one can say that it was probably the respectful aloofness from religious matters that Confucius encouraged which made China an arena of little religious conflict. It may also be suspected that the greater separation from religion explains why Chinese science was able to advance so much further than Greek science, until the emergence of modern science. If not for Eurocentric histories that ignore the contributions of Chinese science and aggrandize the contributions of Greek science, this would have been recognized much sooner.

18. For a discussion of Chinese cosmological ideas, see Graham (1989). Refer especially to the section titled "The Cosmologists," in which he gives a comprehensive account of the classical Chinese correlative cosmology of the *yin-yang* and five-element theories (Graham 1989, pp. 315–370). For a comparative study of Greek and Chinese science, see Lloyd (2004).

19. For an overview of Greek science, refer to Lindberg (1992), Chapters 2–6.

20. It may be supposed that a case could be made to the advantage of the Greeks by arguing that even if the Chinese were theistically agnostic about the gods, the Greeks were methodologically more sophisticated. They may have believed in gods but did not invoke them within their explanations of natural phenomena. The naturalistic orientation of the Presocratics has been seen as a turning point toward scientific thinking. See Wiebe (1991), especially Chapter 2, "Mythopoeic and Scientific Thought."

21. Nevertheless, such a dialogical history need not deny that Greek science made important contributions to modern science. It only makes multicultural contributions mediated through Greek science relevant to answering Needham's Grand Question. See the studies of Diop (1991) and Bernal (1987), both of whom argue that Egyptian (and, in the case of Bernal, also Levantine) science played an important role in the rise of Greek science.

CHAPTER 3

1. Of course there have always been early isolated dissenting voices. For example, Muhammad Iqbal (1934) had argued that Islam gave birth to the "inductive intellect" which led to modern science. He maintained that the reactions of Muslim theologians—like Ibn Hazm and Ibn Taymiyah—against the logic of Aristotle provided the stage for Mill's inductive logic. He also notes that Roger Bacon, often credited with the discovery of the scientific spirit of inquiry, was himself educated in universities in Spain set up by Muslims (p. 23). See Fakhry (2004, pp. 363–368) for a more detailed discussion of Iqbal's views. Nevertheless, none of these dissenting perspectives have been articulated in detail so as to provide a systematic alternative to the Eurocentric interpretations of the rise of modern science. Iqbal's views can also be seen as one-sided because they fail

to see the unique achievements of the moderns as the result of the dialogical integration of not just Arabic but also other traditions.

2. For Kant's view, see the preface to the second edition of his *Critique of Pure Reason* (trans. Kemp, 1950). Much of the discussion here depends on Cohen's excellent exposition of various approaches to the history of science, especially the Scientific Revolution associated with the birth of modern science. However, I have not followed Cohen (1994) in the way he categorizes the approaches of Kant, Whewell, Koyré, Burtt, Dijksterhuis (1961), and others.

3. See Blaut (2000) and Hobson (2004) for a critical discussion of many of these invoked factors. For a typical Marxist position see Bernal (1971).

4. Butterfield (1957) was preceded in this regard by Duhem (1985), and followed by Shapin (1996). Indeed once we stress the importance of medieval contributions as crucial to the emergence of modern science in the seventeenth century, it is easy to reject the discontinuity thesis. Such a discontinuity between modern science and its medieval antecedents might appear to militate against the idea that the medieval contributions were crucial.

5. Of course, it is possible to make a more nuanced argument that sees the medieval contributions as necessary, but not sufficient for the rise of modern science. This would preserve the discontinuity thesis but acknowledge the indispensability of these contributions. This third position is adopted by Needham, who goes further by suggesting that not just medieval European science but also Chinese science made crucial contributions to modern science, but acknowledges discontinuities between the two.

6. See Hetherington (1996).

7. The decline was accompanied by the fading of the influence of Averroism in Europe, although its impact was still felt to some extent in the age of the Enlightenment. See Wahba and Abousenna (1996).

CHAPTER 4

1. Diop pioneered the study of the impact of Egyptian philosophy and science on the Hellenic world more than three decades before Bernal, although he did not examine the Levantine influence on Hellenism. See Diop (1991). There is also now an emerging awareness of the impact of Indian influences on Greek and Hellenistic science. See Goonatilake (1984). It is also likely that such dialogical impacts involved a two-way process in which Indian and Greek ideas influenced each other. See McEvilley (2002).

2. Duhem quoted in Cohen (1994, p. 45).

3. There is no better record of the Persian influence on the Greek world than Herodotus' *Histories* (1998). Surprising as it might appear, most of his history is concerned with the politics and military actions of Persian kings. By contrast Thucydides' history is more concerned with the actions and motives of Greek personalities.

4. Hess also sees wider implications for his dialogical approach to history. It could promote greater awareness of traditional knowledge as a reservoir of resources

for advancing knowledge in the future and exposing the limits of current scientific thinking. He writes:

> By showing that non-Western knowledges, technologies and medicines are often coherent and elaborate, an intellectual resource emerges that can be used to resist the ideology of development interests who wish to impose unwanted Western knowledges and technologies in the name of civilization. Furthermore, by showing that non-Western ways of knowing and doing are often efficacious and in some cases superior to Western or cosmopolitan alternatives, it is possible to build a resource base for critiquing and contributing to changing development projects .... At another level, the study of ethno-knowledges makes it possible to put into question the universalistic assumptions of cosmopolitan science and technology. (Hess 1995, p. 210)

5. For a more detailed historical discussion of the origin and spread of these technologies, see Pacy (1990) and Hobson (2004).
6. We will be using the Latinized name Averroes, instead of the Arabic name Ibn Rushd, since this was the name by which he was better known in the Western world, and all discussion of his views in this study will be focused on his influence on European thought.
7. We will be using the name Alhazen in this study because we are concerned with Ibn al-Haytham's impact on European thought in the medieval and early modern era, when he was better known by his Latinized name.
8. The discoveries of these historical parallels were made by a chain of historians beginning with Victor Roberts in 1957, followed by Kennedy, Neugebauer, Swerdlow, and Saliba. See Saliba (1994, p. 254). See also King (2000).

CHAPTER 5

1. Needham (1958), e.g., argues that Chinese cosmological ideas transmitted by Jesuit astronomers in China had an impact on the development of theoretical cosmology in early modern Europe. This influence is addressed in greater detail later in this study.
2. Goonatilake, "A Project for Our Times," in Sardar (1988) p. 234.
3. For an alternative approach that adopts a different criterion to establish transmission, see Aryabhata Group (2002). They provide an approach based on legal criteria used to determine culpability in murder cases—motivation, opportunity, circumstantial evidence, and direct evidence—to establish transmissions.

CHAPTER 6

1. W. G. de Burgh (1953) emphasizes the close parallels between Judaic, Christian, and Islamic traditions in their attempts to reconcile science and religion:

> Despite their diversity and antipathies, these three religions had much in common. All three were monotheist, and taught the doctrine of a divine

creation and providential government of the world, of a supernatural sphere peopled by supernatural powers, of heaven and hell, of the immortality of the soul and the hope of personal salvation. All three appealed to a revelation embodied in sacred writing and to the authority of prophetic teachers— Moses, Jesus, Mohammed—who had delivered an inspired message to mankind. All three found themselves obliged to explain and defend their faith in the face of heretics and unbelievers, and had recourse to the allegorical method, first utilized by the Stoics, as the instrument of theological exegesis. In other words, all alike felt, though in varying measure, the need of a rational theology, which should harmonize science and revelation .... It is the community of aim and method among the three streams of thought that gives the philosophy of the Middle Ages its characteristic unity. (vol. 2, p. 447)

2. Maimonides (1956) began the work under the inspiration of Aristotelianism and as an attempt to rebut the views of Islamic *kalam*. The work itself was originally written in Arabic but was translated into Latin and had a profound influence on medieval European philosophy. In his study there is an extended discussion of both the *kalam* and the *falsafah* schools of Arabic philosophy, so that Europeans were able to use it to take the scholastic tradition off from the point in the debate and dialogue already reached by Arabic thinkers. See Burrell (1993); Wahba and Abousenna (1996).

3. For a more extensive discussion of the impact of the plague on trade and life in the medieval world of China, Islam, and Europe, see Abu-Lughod (1989).

4. It is noteworthy that the Mohists in China also developed many ideas that closely parallel modern ideas. They emphasized universal love of mankind (as Enlightenment philosophies did), cultivated studies in logic, and developed utilitarian conceptions of nature. This philosophy sprang up among the craftsmen of China, who were precisely the elements of Chinese society concerned with technology. Although Mohism was a strong contender against Confucianism in the period of the warring states, it subsequently declined. However, what is significant is that as a result of their close relationship and involvement with mechanical technologies, Mohists were led to views that, in many respects, parallel those that developed in Europe in the seventeenth century. Even more striking is the fact that the Mohists were the only school in China to propose atomic ideas. This lends credibility to the notion that technology is a carrier of orientations toward nature, and puts into question White's assumption that technology and science are easily separable. See Chapter 2 in Graham (1989) for a discussion of Mohist views and the artisan traditions that influenced it. Also see Graham (1978) for a more detailed study of Mohism.

5. See Marco Polo's *The Travels* (1958) for details of the sort of employment he, and others from the European and Arabic worlds, managed to procure within the Mongol Empire.

6. The great religious toleration practiced by the Mongols was noted by Marco Polo. He quotes Kublai Khan:

There are four prophets who are worshipped and to whom all the world does reverence. The Christians say that their God was Jesus Christ, the

Saracens Mahomet, the Jews Moses and the idolaters (i.e. Buddhists) Sakyamuni Burkhan, who was the first to be represented as God in the form of an idol. And I do honor and reverence to all four, so that I may be sure of doing it to him who is greatest in heaven and truest; and to him I pray for aid. (Marco Polo 1958, p. 119)

7. The search for Prester John perhaps began with Marco Polo's account of his travels, in which he argues that descendents of a great king Prester John, who once ruled the Mongols but were overthrown by Genghis Khan, continue to rule the territory of Tenduc (a little west of China) as vassals of the Mongol Khan. Marco Polo described the ruling house of Tenduc as Christian although there were many Muslims and Buddhists among its subjects (Marco Polo 1958, pp. 92–107).

8. Joseph (2000, p. 315). See also Goldstein (1980, p. 127).

9. According to Goldstein, al-Bitruji's purely Aristotelian system was incompatible with Ptolemaic astronomical theory and generated controversy until the seventeenth century (Goldstein 1980, p. 112). However, as we will see later, the Maragha School, which rejected the Aristotelian model, played a more important role in the modern scientific revolution.

10. Bakar (1999, pp. 152–154); Goldstein (1980, pp. 124–126). For a more detailed discussion of this Averroist influence from the fourteenth to the seventeenth centuries, see Wahba and Abousenna (1996).

11. For a historical overview of these influences, see Smart (2000).

12. Translated and quoted in Fakhry (2004, p. 71).

13. Translated and quoted in Fakhry (2004, p. 71).

14. Hobson (2004, p. 174) has described this as maintaining a custodial or librarian view of Arabic thinkers as holders and translators of ancient Greek texts who did no original work.

15. St. Augustine, the leading influence on medieval Christian views, advised the faithful thus:

> When, then, the question is asked what we are to believe in regard to religion, it is not necessary to probe into the nature of things, as was done by those whom the Greeks called *physici;* nor need we be in alarm lest the Christian should be ignorant of the force and number of the elements— the motion, and order, and eclipses of the heavenly bodies; the form of the heavens; the species and the natures of animals, plants, stones, fountains, rivers, mountains, about chronology and distances; the signs of coming storms; and a thousand other things which those philosophers either have found out, or think they have found out ... It is enough for the Christian to believe that the only cause of all created things, whether heavenly or earthly, whether visible or invisible, is the goodness of the Creator, the one true God; and that nothing exists but Himself that does not derive its existence from him. (Quoted in Kuhn [1957, p. 107] from *St. Augustine's Works,* edited by Marcus Dods [Edinburgh: Clark, 1871–1877], lx, 180–181)

It is interesting to compare this Augustinian view with al-Ghazali's admonition to youth interested in scientific studies we quote in Chapter 9.

CHAPTER 7

1. Kuhn's study, although made nearly half a century ago, and despite many studies of the Copernican and the Scientific Revolution since then, remains a classical text. Internalist intellectualist histories went out of fashion not only because Kuhn did such a thorough job with his study but also because his more influential later work *Structure of Scientific Revolutions* (Kuhn 1970) made externalist approaches fashionable. Hence Kuhn's study is doubly significant—it brings to fruition the internalist approaches pioneered by earlier historians and paves the way for the externalist approaches that came later. Sharrock and Read (2002) also argue that it served as an important case study for his concept of scientific revolutions as mediated by paradigm changes that profoundly altered our understanding of the history, philosophy, and sociology of science.

2. Dijksterhuis (1961, pp. 288–289) continues:

   > After a stagnation of about fourteen centuries the evolution of astronomy continued at Frauenburg [where Copernicus resided] at the point where it left off in Alexandria ... he considered the greatest gain it had brought astronomy was not the changed position of the sun in the universe and the resulting simplification of the world-picture, but the abolition of the *punctum aequans,* the atonement for the sin against the spirit of Platonic philosophy which Ptolemy had committed in an evil hour.

   The spirit of Platonic philosophy is clearly expressed in the sixth century CE by Simplicius when he writes in his commentary on Aristotle's *De caelo*:

   > Plato lays down the principle that the heavenly bodies' motion is circular, uniform, and constantly regular. Thereupon he sets mathematicians the following problem: what circular motions, uniform and perfectly regular, are to be admitted as hypotheses so that it might be possible to save the appearances presented by the planets. (Quoted in Hetherington 1996, p. 271)

3. One of the earliest to take Aryabhata's achievements further was Varahamihira (c. 505–587), who extended his trigonometric discoveries by discussing sine, cosine, and versine relations. See Joseph (2000, pp. 266–267).

4. In the ninth century Mahavira (c. 850), who came from the nonastronomical tradition of Jain mathematics, was to fuse Jain techniques with the Hindu mathematical-astronomical tradition. Mahavira's studies came to be known to the Arabic scholars through translations. His contributions included finding solutions to different types of quadratic equations, advancing geometrical studies of right-angled triangles whose sides are rational, and even attempting to derive general mathematical formulae for determining the area and perimeter of an ellipse. Although the problem was not solved by him, it opened the door to later lines of research concerned with determining areas and perimeters of figures not defined by straight lines or circles. (Joseph 2000, pp. 267–268).

5. The discussion that follows is based on a paper coauthored by Dennis Almeida and George Gheverghese Joseph under the name Aryabhata Group (2002). It is part of a project to determine whether the discoveries of the Kerala School of mathematics influenced mathematical thought in modern Europe—especially in relation to the development of infinitesimal calculus.

6. It is also noteworthy that the method used by Fermat in 1629 to evaluate areas under parabolas of the form $y = x^k$, is also described by Jyesthadeva in the *Yuktibhasa*, and was formulated in the *Tantrasangraha* of his predecessor Nilakantha. Although credited in Europe as a discovery of Fermat, it was also used by Roberval and Cavalieri, and adopted by Wallis to obtain a value for π. Joseph and Almeida note that many European mathematicians suddenly happened to have "discovered" the method at about the same time, though it did not have an epistemological basis in European mathematics as it did in its Indian counterpart. They cite it as one more item of circumstantial evidence that Indian mathematical discoveries of the Kerala School reached Europe. (Aryabhata Group 2000, pp. 42–43)

7. There is also another major consequence of these intellectual exchanges that has yet to be systematically studied. It brought together the epistemological orientation of Indian mathematics with the epistemological orientation of Greek mathematics—a volatile mixture that came to be accommodated into modern mathematics only after more than 200 years of controversy and only at the end of the nineteenth century. In the *Crest of the Peacock* Joseph argues that Indian mathematics had a different epistemological basis from Greek mathematics, being inspired by the Paninian empiricist orientation rather than the Greek rationalist approach of Plato. In his treatise *Astadhyayi* Panini gave rules that he empirically established for properly generating compound nouns and sentences by operating on underlying elements made up of nouns, verbs, vowels, and consonants. Joseph describes Panini's achievement as follows:

> [Sanskrit's] potential for scientific use was greatly enhanced as a result of the thorough systematization of its grammar by Panini ... On the basis of just under 4000 *sutras* (i.e. rules expressed as aphorisms), he built virtually the whole structure of the Sanskrit language, whose general "shape" hardly changed for the next two thousand years ... An indirect consequence of Panini's efforts to increase the linguistic facility of Sanskrit soon became apparent in the character of scientific and mathematical literature. This may be brought out by comparing the grammar of Sanskrit with the geometry of Euclid—a particularly apposite comparison since, whereas mathematics grew out of philosophy in ancient Greece, it was ... partly an outcome of linguistic developments in India. (Joseph 2000, p. 217)

One important result of Paninian linguistics was that it made it possible to establish in India a phonetically based alphabet system of great flexibility that enabled words to be written as compounded out of letters that represented sound elements. It continues to constitute the main approach to Indian alphabet systems until the present time. Another consequence of the phonetic alphabet system was that it facilitated the emergence of first a word numeral system, and later an alphabet system, that made it possible to represent large numbers. This point has been stressed by Joseph:

> [The word numeral system] led to the adoption of a series of names for successive powers of ten. The importance of these number-names in the evolution of the decimal place-value notation cannot be exaggerated. The word-numeral system, later replaced by an alphabetic notation, was the logical outcome of proceeding by multiples of ten. Thus 60799 is *sastim* (60) *sahsara* (thousand) *sapta* (seven) *satani* (hundred) *navatim* (nine ten times) *nava* (nine). (Joseph 2000, p. 242)

The underlying principle in the number system, as in the linguistic system of Panini, is that in both cases words and numbers are seen as compounded of more primitive atomic elements to form a whole—e.g., a word is written by a combination of symbols representing sounds; a number, as a combination of symbols representing "atomic" numbers with absolute and position value. It would be fruitful to examine whether Paninian linguistic atomism influenced Indian mathematical atomism and Indian physical atomism. This would explain the common atomic mode of analysis that underlies Indian linguistics, mathematics, and theories of matter adopted by most schools. It is a question that merits deeper investigation.

8. These criticisms also influenced medieval European thinkers such as Albertus Magnus, Robert Grosseteste, and Roger Bacon (Ronan 1983, p. 218).
9. Translated and quoted in Nasr (1968, p. 177).

CHAPTER 8

1. For a more detailed examination of Alhazen's experimental method and its influence, see Omar (1997).
2. This interpretation is supported by Goldstein:

> [Alhazen] became the prime source of optical knowledge for the European Middle Ages and Renaissance. Men like Roger Bacon, Leonardo da Vinci, Johannes Kepler were inspired by his insights and influenced by his methodological approach. If Islam was teaching Medieval Europe to see, Alhazen taught the most incisive lesson in visual precision; no wonder the scientist-artist Leonardo da Vinci was his particular admirer. (Goldstein 1980, p. 116)

3. However, it is important to note that although Hanson, Toulmin, Kuhn, and Feyerabend repudiate the visual or ocularcentric paradigm that observation constitutes an incorrigible foundation for knowledge, none of them repudiates the importance of testing theories against observation.

CHAPTER 9

1. For example, the historian Westfall (1977) assumes that the atomic view of the ancients reappeared in modern Europe only as a result of the recovery of the Hellenistic heritage, and soon came to be seen by many thinkers as compatible with, or even identical to, the mechanical worldview. He writes:

> Inevitably, the atomic philosophy of antiquity had reappeared in Western Europe in the general recovery of ancient thought during the Renaissance. Galileo had felt its influence, and its mechanistic treatment of nature probably helped to shape Descartes' systems. It remained, however, for a contemporary of Descartes, Pierre Gassendi (1592–1655) to espouse and expand atomism as an alternative mechanical philosophy. (Westfall 1977, p. 39)

2. For a more detailed exposition of Islamic *kalam* see Fakhry (1958); also Harry A. Wolfson (1976). Maimonides in his *Guide to the Perplexed* also offers an extensive overview, as well as critique, of *kalam*.
3. Bakar argues that since Maimonides was translated into Latin around 1220 CE, his work could have provided the basis for Aquinas' critique of Islamic occasionalism (Bakar 1999, pp. 100–101).
4. See Goldstein (1980, pp. 125–126).
5. Quoted in Hoodbhoy (1991, p. 107) from al-Ghazali, *Ayyuha-al-Walad*, trans. G. H. Scherer (1932), Beirut: American Press, p. 57. Although al-Ghazali is often charged with causing the decline of Arabic science, it is possible to suspect this conclusion because Arabic science developed in ways that broke away from the Hellenistic tradition with Alhazen optics and Maragha School astronomy. It is probably more accurate to see al-Ghazali as causing the decline of the Aristotelian and rationalist schools of science, but not science per se. As we will see later, his epistemological critique was an important stage in the demise of Greek science and the emergence of modern science.
6. Quoted in Bakar (1999, p. 92). Bakar's quote is taken from de Berg's translation of Averroes' *Tahafut al-Tahafut* in E. J. Gibb Memorial Series, New Series 19 (London: Luzac and Co., 1954), pp. 316–317.
7. Quoted in Bakar (1999, p. 41).
8. Newton (1952, pp. 400–401). Quoted in Kuhn (1957, p. 260).
9. See Osler (1994) for a more detailed study of these issues.

CHAPTER 10

1. The importance of Jundishapur in bringing together knowledge originating in the Hellenistic world, China, and India before the arrival of Islam merits more attention than it has hitherto received. It was founded by Shapur I (241–272) after he defeated the Roman army in 260 CE. The closure of Plato's academy in 529 CE forced many scholars to seek refuge in Jundishapur, and the burning and closure of the library at Alexandria also led to a new influx of scholars.

The Kushan Empire, which was an important area from which Buddhism entered China, also led Jundishapur to flourish through the interaction of Chinese ideas with Hellenistic and Indian thought. As a result of their conquests the Arabic Muslims inherited the scholastic ecumenism of the Kushan Empire in general, and Jundishapur in particular. See Joseph (2000, pp. 9–22) for a more detailed survey of these cross-cultural interactions.

2. In his study *Parallel Developments* the Japanese comparative philosopher and historian Hajime Nakamura writes:

> In the West the two terms [religion and philosophy] have been fairly sharply distinguished from each other, while in Eastern traditions the dividing line is often difficult to discern. If we insist on being too strict in our definitions, we fail to catch many common problems. It is possible that an idea or attitude held by a Western philosopher finds its counterpart not in an Eastern philosopher but in an Eastern religious thinker and vice versa. (Nakamura 1975, p. 3)

3. One important factor that made it possible for the Arab *falsafah* tradition to be so easily naturalized in the Western medieval world may be attributed to the influence of the *Liber de Causis* of Proclus, which became an authoritative work among European scholastics. It drew commentaries by Thomas Aquinas and Albert the Great. Like most so-called neo-Platonic works it saw the universe as an emanation of a single principle that medieval scholastics could identify with God. Of course, like the Arabic theological thinkers, Aquinas was himself to reject this emanation theory since it did not accord with the creation of the world ex nihilo by God.

4. It is also significant that after al-Ghazali, philosophical development in the Arabic-Islamic world begins to assign more significance to mystical experience and illumination than reason, and come to have greater affinity to Indian than Hellenic philosophical orientations. For survey accounts of the views of these later philosophers, see McGreal (1995). See also Fakhry (2004), especially Chapter 10, "Post-Avicennian Developments: Illumination and Reaction Against Peripateticism."

5. See also Bose, Sen, and Subbarayappa (1971, p. 318), who argue for the possibility of interaction of the Arabic and Indian alchemical traditions, especially in the context of the confluence of the two medical traditions.

6. Descartes, *Selections,* p. 15 (Ralph M. Eaton, ed., 1955).

CHAPTER 11

1. Ronan's *Shorter Science and Civilization in China* is an abridgement of Needham's views. It will be assumed, where quoted, unless otherwise specified, that Ronan faithfully represents Needham's views. This seems reasonable since the work was written in close discussion with, and the approval of, Needham.

2. Duhem trans. and quoted in Cohen (1994, p. 52).

3. It is also important to remember that the scholastic thinkers could not have been unaware of the ideas of al-Ghazali either directly, or through attacks on them by Maimonides and Averroes, both of whom wrote in the context of debates that took place in the Arabic-Islamic cultural arena. See Fakhry (1958), Hanley (1982), Burrell (1988), Bullough (1996), Leaman (1996), Wahba (1996), and McAleer (1999).
4. For a translation of Harvey's original Latin text in English see Harvey (1963).
5. For a study of the contrasting logic of Greek and Chinese thought see Shankman and Durrant (2002) and Lloyd and Sivin (2002).
6. For Kant's view, see the preface to the second edition of his *Critique of Pure Reason* (trans. Kemp, 1950).

CHAPTER 12

1. For a more detailed discussion of the Chinese pole star-based astronomy see Chapter 2, Ronan (1981). The differences in the astronomical traditions even led to some misunderstanding because Chinese astronomy depended on the observation of stars near the pole and the equator, whereas European (and Greek) astronomy measured motion with reference to the path of the sun across the constellations of the zodiac. Some of these misconceptions were not cleared up until the end of the nineteenth century, by which time the Chinese approach to observations had become the dominant one even in Europe (Ronan 1981, p. 68).
2. Ronan (1981, p. 213). Quotes from Needham (1958, p. 2).
3. See Sun (2000) for a more detailed discussion of these models, especially pp. 437–443. See also Ho (2003).
4. Sun (2000, p. 441).
5. Sun argues that this theory appears the most philosophically sophisticated of the three theories although the least capable of serving the goal of yielding a mathematical model of the universe, since it assumed that the heavenly bodies moved erratically (Sun 2000, p. 438).
6. Also quoted in Ronan (1981, pp. 214–215).
7. In recent years the literature on Jesuit missions to China and the rest of the world has literally exploded. It is now increasingly recognized that the Jesuits played a role not only in spreading religion abroad, but also in communicating ideas from the wider world to a receptive Europe. The following is just a sampling of this literature: Anagnostou (2002), Chapman (1984), Correia-Afonso (1997), Donnelly (1982), Feingold (2003), Harris (1996), Needham (1958), O'Malley et al. (2005), Udías (1994), and Wallace (1991).
8. See Franke (1967) for a history of these Chinese connections with the West.
9. Kuhn (1970) has emphasized the importance of paradigms in shaping scientific observations. He explains why European astronomers failed to see comets as astronomical data. He argues that their paradigmatic belief in an unchanging heavens precluded the appearance of erratic events, such as comets.
10. This does not preclude the possibility that further studies may reveal documentary evidence to show that European astronomers were directly considering the

views of Chinese thinkers transmitted back to them in written communiqués, possibly by Jesuit missionary-scientists. Such evidence would only make the case for dialogical exchanges between Chinese and modern European science stronger.

11. For a more detailed discussion of these Chinese discoveries see Ronan (1983), especially Chapter 2 on Chinese science.

CHAPTER 13

1. Philolaus, the Pythagorean, was another ancient astronomer who considered the earth to move, but his universe was not heliocentric. He considered the heavenly bodies to revolve around a central fire in ten concentric circles—with the fixed stars on the outermost circle, followed by the five planets, the sun, the moon, the earth, and a mysterious counter-earth in that order.

2. Kuhn writes:

> The conception of a planetary earth was the first successful break with a constitutional element of the ancient world view. Though intended solely as an astronomical reform, it had destructive consequences which could be resolved only within a new fabric of thought. Copernicus himself did not supply that fabric; his own conception of the universe was closer to Aristotle's than to Newton's. But the new problems and suggestions that derived from his innovation are the most prominent landmarks in the development of the new universe, which that innovation itself had called forth. (Kuhn 1957, p. 264)

3. For a critique of Kuhn's Eurocentrism see Bajaj (1988). Surprisingly Bajaj criticizes even Needham on this score. Needham argues that Newtonian science was made possible by the Western belief in a monotheistic deity who defined the universal laws of nature. Hence the Chinese, with their organic materialist worldview, could not have created Newtonian science. Moreover, Needham maintained that although modern science, with the quantum theory, has gone beyond Newtonian science toward a view closer to Chinese organic materialism, the Chinese could not have discovered quantum theory since Newtonian science constituted a necessary intermediate stage to create it. Bajaj sees Needham as asserting the Eurocentric claim that "not only are the final truths arrived at by Western science uniquely valid, but the exact historical sequence through which they were arrived at in the West was essential and necessary." (Bajaj 1988, pp. 59–60)

4. Descartes (1955), p. 16.

5. Descartes also writes: "And for this purpose it was requisite that I should borrow all that is best in Geometrical Analysis and Algebra, and correct the errors of the one by the other" Descartes (1955), p. 18.

6 Descartes (1955), p. 19.

7. See also Keynes (1947, pp. 27–34).

8. Martin Bernal argues that ignoring the influence of Egyptian ideas on the Scientific Revolution marginalizes the African influence on modern science.

See Bernal (1987), pp. 151–169 for the influence of the Hermetic tradition on many of the pioneers of modern science from Copernicus to Newton.

CHAPTER 14

1. The term "Axial Age" was used by Karl Jaspers to refer to the period between the eighth and second centuries BCE, in which many of the great philosophies and religions of the world arose. It includes the lifetimes of Confucius and Lao-tzu in China, the period in which the Upanishads were produced and Buddha lived in India, the time of Zoroaster in Persia, the epoch of the great prophets of Israel, and the era of Plato and Aristotle in Greece. For more about this seminal period see Eisenstadt (1986), Burkert (2004), Arnason, Eisenstadt, and Wittrock (2004).

2. Indeed Bacon, Descartes, Galileo, and many others were acutely conscious of the novelty of the science they were creating and its break with the ancient Greek heritage. This sense of dichotomy continued with the *philosophes* in the era of the Enlightenment and persisted into the nineteenth century. It was only in the Romantic age that there developed a sense of continuity of the modern tradition with the Greek tradition of science. See Baumer (1977).

3. The dialogical answer also puts to rest the oft-made charges that Needham's Grand Question is intrinsically Eurocentric, illegitimate, or irrelevant. Hobson suggests that scholarship errs when it follows Weber to ask what made it possible for the West to break through into modernity and what prevented the East from doing so. According to him this naturally promotes attempts to find immanent factors in Europe and elsewhere to account for the success in one case and failure in the other (Hobson 2004, p. 295). The sinologist Graham thinks that although it is reasonable to consider what factors brought about an event like the Scientific Revolution, it is an illegitimate pseudoproblem to inquire why the same factors did not come together elsewhere (Graham 1989, p. 317). Sivin makes the same point more graphically by means of an analogy: although it is legitimate to ask why your name appeared on page 3 of the newspaper today, it is absurd to ask why it did not (Sivin 1984, p. 536). The historian Saliba considers attempts to explain why modern science did not develop elsewhere to be irrelevant. It is more important to understand another tradition in its own right, within its own context, in terms of the research programs that inspire it (Saliba 1994, p. 29).

However, the dialogical answer this study defends shows that Needham's Grand Question is not necessarily Eurocentric, illegitimate, or irrelevant. It only becomes so if it is conjoined to the assumption that answers to the Grand Question must be in terms of immanent factors within Europe, and other civilizations.

Bibliography

Abu-Lughod, Janet L. (1989). *Before European Hegemony: The World System A.D. 1250–1350*. Oxford: Oxford University Press.

Abu-Shanab, Robert E. "Ghazali and Aquinas on Causation." *Monist* (1974) 58: 140–150.

Adamson, Peter and Richard C. Taylor (eds.) (2005). *The Cambridge Companion to Arabic Philosophy*. Cambridge: Cambridge University Press.

Al-Andalusi, Said (1991). *Science in the Medieval World: "Book of the Categories of Nations."* Trans. Semaan I. Salem and Alok Kumar. Austin: University of Texas Press.

Al-Ghazali, Abu Hamid Muhammad (2002). *The Incoherence of the Philosophers*. Ed. and trans. Michael E. Marmura. Provo, Utah: Brigham Young University.

Alberuni (1983). *India: An Account of the Religion, Philosophy, Literature, Geography, Chronology, Astronomy, Customs, Laws and Astrology about AD 1030*. Vols. 1 and 2. Trans. Edward C. Sachau. New Delhi: Oriental Books Reprint Corporation.

Alhazen (2001). *Alhacen's Theory of Visual Perception: A Critical Edition*. Trans. and ed. Mark Smith. Philadelphia: American Philosophical Society.

Alvares, Claude Alphonso (1979). *Homo Faber: Technology and Culture in India, China and the West 1500–1972*. New Delhi: Allied Publishers.

Amin, Samir (1989). *Eurocentrism*. Trans. Russell Moore. New York: Monthly Review Press.

Anagnostou, Sabine. "Jesuit Missionaries in Spanish America and the Transfer of Medical-Pharmaceutical Knowledge." *Archives internationales d'histoire des sciences* (2002) 52(148): 176–197.

Ansari, S. M. Razaullah (ed.) (2002). *History of Oriental Astronomy*. Dordrecht, Netherlands: Kluwer.

Arnason, Johann P., S. N. Eisenstadt, and Bjorn Wittrock (eds.) (2004). *Axial Civilizations and World History*. Leiden, Boston: Brill Academic Publishers.

Aryabhata Group (2002). "Transmission of Calculus from Kerala to Europe." In Aryabhateeyam Proceedings (2002). pp. 33–48.

Aryabhateeyam Proceedings (2002). Proceedings of the International Seminar and Colloquium on 1500 Years of Aryabhateeyam. Kochi, India: Kerala Sastra Sahitya Parishad.

Ashman, Keith M. and Philip S. Baringer (eds.) (2001). *After the Science Wars*. London and New York: Routledge.

Averroes (1954). *The Incoherence of the Incoherence.* Trans. Simon Van Den Bergh. London: Luzac.

Baber, Zaheer (1996). *The Science of Empire: Scientific Knowledge, Civilization and Colonial Rule in India.* Albany, NY: State University of New York Press.

Baigrie, Brian S. (2001). *The Renaissance and the Scientific Revolution: Biographical Portraits.* New York: Charles Scribner's Sons.

Bajaj, Jatinder K. "Francis Bacon, the First Philosopher of Modern Science: A Non-Western View." In Nandy (1988).

Bakar, Osman (1998). *Classification of Knowledge in Islam: A Study in Islamic Philosophies of Science.* Cambridge, UK: Islamic Texts Society.

——— (1999). *The History and Philosophy of Islamic Science.* Cambridge, UK: Islamic Texts Society.

Basham, A. L. (ed.) (1975). *A Cultural History of India.* Oxford: Clarendon Press.

Baumer, Franklin L. (1977). *Modern European Thought: Continuity and Change in Ideas, 1600–1950.* New York: Macmillan.

Berman, Morris (1984). *The Reenchantment of the World.* New York: Bantam Books.

Bernal, J. D. (1971). *Science in History.* Cambridge, MA: MIT Press.

Bernal, Martin (1987). *Black Athena: The Afroasiatic Roots of Classical Civilization.* New Brunswick: Rutgers University Press.

Blaut, James M. (1993). *The Colonizers Model of the World: Geographical Diffusionism and Eurocentric History.* New York: Guilford Press.

——— (2000). *Eight Eurocentric Historians.* New York: Guilford Press.

Bodde, D. (1991). *Chinese Thought, Society and Science: The Intellectual and Social Background of Science and Technology in Pre-Modern China.* Honolulu: University Press of Hawaii.

Bose, D. M., S. N. Sen, and B. V. Subbarayappa (1971). *A Concise History of Science in India.* New Delhi: Indian National Science Academy.

Bronkhorst, Johannes. "Panini and Euclid: Reflections on Indian Geometry." *Journal of Indian Philosophy* (2001) 29(1–2): 43–80.

Brown, James R. (2001). *Who Rules in Science? An Opinionated Guide to the Wars.* Cambridge, MA: Harvard University Press.

Bullough, V. L. "Medieval Scholasticism and Averroism: The Implication of the Writings of Ibn Rushd to Western Science." In Wahba and Abousenna (1996), pp. 41–51.

Burkert, Walter (2004). *Babylon, Memphis, Persepolis: Eastern Contexts of Greek Culture.* Cambridge, MA: Harvard University Press.

Burrell, David (1988). "Aquinas's Debt to Maimonides." In *A Straight Path: Studies in Medieval Philosophy and Culture* (Essays in Honour of Arthur Hyman). Ed. Ruth Link-Salinger. Washington: Catholic University of America Press.

——— (1993). "Aquinas and Islamic and Jewish Thinkers." In *The Cambridge Companion to Aquinas.* Ed. Norman Kretzmann. New York: Cambridge University Press.

Burtt, E. A. (1959). *The Metaphysical Foundations of Modern Physical Science: A Historical and Critical Essay.* London: Routledge & Kegan Paul.

Butterfield, Herbert (1957). *The Origins of Modern Science, 1300–1800.* New York: The Free Press.

Carrier, Martin, Johannes Roggenhofer, Gunter Knppers and Philippe Blanchard (eds.) (2004). *Knowledge and the World: Challenges Beyond the Science Wars.* New York: Springer.

Carroll, William. "Creation, Evolution and Thomas Aquinas." *Revue des Questions Scientifiques* (2000) 171(4): 319–347.

Chan, Wing-Tsit (1969). *A Source Book in Chinese Philosophy.* Princeton, NJ: Princeton University Press.

Chapman, Allan. "Tycho Brahe in China: The Jesuit Mission to Peking and the Iconography of European Instrument-Making Processes." *Annals of Science* (1984) 41: 417–443.

Chattopadhyaya, Debiprasad (1996). *Interdisciplinary Studies in Science, Technology, Philosophy and Culture.* New Delhi: Project of History of Indian Science, Philosophy, and Culture.

Chau, P. L. "Ancient Chinese Had Their Finger on the Pulse." *Nature* (2000) 404(6777): 431.

Cheng, Tsung O. "Did Greeks Beat Chinese on Blood Circulation?" *Nature* (2000) 405(6790): 993.

Clarke, J. J. (1997). *Oriental Enlightenment: The Encounter between Asian and Western Thought.* London: Routledge.

Cohen, H. Floris (1994). *The Scientific Revolution: A Historiographical Inquiry.* Chicago: University of Chicago Press.

Copernicus, Nicholas (1995). *On the Revolutions of Heavenly Spheres.* Trans. Charles Glenn Wallis. New York: Prometheus Books.

Correia-Afonso, John (1997). *The Jesuits in India, 1542–1773: A Short History* (Studies in Indian History and Culture of the Heras Institute). Anand, Gujarat, India: Gujarat Sahitya Prakash.

Crombie, A. C. (1979). *Augustine to Galileo.* Cambridge, MA: Harvard University Press.

Crowe, Michael J. (2001). *Theories of the World from Antiquity to the Copernican Revolution.* New York: Dover Publications.

Cullen, Christopher (1996). *Astronomy and Mathematics in Ancient China: The Zhou Bi Suan Jing.* Cambridge: Cambridge University Press.

Damerow, Peter (2004). *Exploring the Limits of Preclassical Mechanics: A Study of Conceptual Development in Early Modern Science.* New York: Springer.

Dantzig, Tobias (1954). *Number, the Language of Science: A Critical Survey Written for the Cultured Non-Mathematician.* New York: Macmillan.

Dear, Peter Robert (1995). *Discipline and Experience: The Mathematical Way in the Scientific Revolution.* Chicago: University of Chicago Press.

De Burgh, William George (1953). *The Legacy of the Ancient World.* Melbourne: Penguin Books.

Descartes, Rene (1955). *Selections.* Ed. Ralph M. Eaton. New York: Charles Scribner's Sons.

Dharampal (1971). *Indian Science and Technology in the Eighteenth Century: Some Contemporary European Accounts.* Delhi, India: Impex.

Dijksterhuis, E. J. (1961). *The Mechanization of the World Picture.* Oxford: Oxford University Press.

Diop, Cheikh Anta (1991). *Civilization or Barbarism: An Authentic Anthropology.* New York: Lawrence Hill Books.

Donnelly, John Patrick. "The Jesuit College at Padua: Growth, Suppression, Attempts at Restoration 1552–1606." *Archivum Historicum Societatis Jesu* (1982) 51: 47–79.

Drake, Stillman. "Impetus Theory Reappraised." *Journal of the History of Ideas* (1975) 36: 27–46.

Druart, Therese-Anne (2003). "Algazali." In *A Companion to Philosophy in the Middle Ages.* Ed. Jorg Gracia. Malden, MA: Blackwell Publishing, pp. 118–126.

Duhem, Pierre (1985). *Medieval Cosmology: Theories of Infinity, Place, Time, Void and the Plurality of Worlds.* Trans. R. Ariew. Chicago: University of Chicago Press.

Dunne, George H. (1962). *Generation of Giants: The Story of the Jesuits in China in the Last Decades of the Ming Dynasty.* Notre Dame, Ind.: University of Notre Dame Press.

Eisenstadt, S. N. (1986). *The Origins and Diversity of Axial Age Civilizations.* Albany: SUNY Press.

Emeagwali, Gloria Thomas (ed.) (2002). *Africa and the Academy: Challenging Hegemonic Discourses on Africa.* Trent, NJ: Africa World Press.

Fakhry, Majid (1958). *Islamic Occasionalism and Its Critique by Averroes and Aquinas.* London: Allen and Unwin.

——— (2004). *A History of Islamic Philosophy.* New York: Columbia University Press.

Farrington, Benjamin (1949). *Greek Science : Its Meaning for Us.* New York: Penguin.

Feingold, Mordechai (ed.) (2003). *Jesuit Science and the Republic of Letters.* Cambridge, MA: MIT Press.

Feyerabend, Paul K. (1975). *Against Method: Outline of an Anarchistic Theory of Knowledge.* London: Verso.

Fowden, Garth (1986). *The Egyptian Hermes: A Historical Approach to the Late Pagan Mind.* Cambridge: Cambridge University Press.

Frank, Andre Gunder (1998). *Reorient: Global Economy in the Asian Age.* Berkeley: University of California Press.

Franke, Wolfgang (1967). *China and the West.* Trans. R. A. Wilson. Columbia: University of South Carolina Press.

Fuchs, Thomas (2001). *The Mechanization of the Heart: Harvey and Descartes.* Trans. Marjorie Grene. Rochester, NY: University of Rochester Press.

Fukuyama, Francis (1992). *The End of History and the Last Man.* New York: Avon Books.

Gates, Henry Louis, Jr. (ed.) (1992). *Loose Canons: Notes on the Culture Wars.* Oxford: Oxford University Press.

Gillespie, Charles C. (1960). *The Edge of Objectivity: An Essay in the History of Scientific Ideas.* Princeton, NJ: Princeton University Press.

Gingerich, Owen (2004). *The Book Nobody Read: Chasing the Revolutions of Nicolaus Copernicus.* New York: Walker & Company.

Girish, T. E. and Radhakrishnan Nair (2002) "On the Physical Basis of Indian Geometrical Ideas on Planetary Motion." In Aryabhateeyam Proceedings (2002) pp. 83–92.

Goldstein, Thomas (1980). *Dawn of Modern Science: From the Arabs to Leonardo da Vinci.* Boston: Houghton Mifflin.

Goody, Jack (1996). *The East in the West.* Cambridge: Cambridge University Press.

Goonatilake, Susantha (1984). *Aborted Discovery: Science and Creativity in the Third World.* London: Zed Books.

———— "A Project for Our Times." In Sardar (1988), pp. 226–238.

Graham, A. C. (1978). *Later Mohist Logic, Ethics and Science.* Hong Kong: Chinese University Press.

———— (1989). *Disputers of the Tao: Philosophical Argument in Ancient China.* La Salle, IL: Open Court.

Grant, Edward (1996). *The Foundations of Modern Science in the Middle Ages.* Cambridge: Cambridge University Press.

Gross, Paul R., and Norman Levitt (1994). *Higher Superstition: The Academic Left and Its Quarrels with Science.* Baltimore and London: Johns Hopkins University Press.

Gross, Paul R., Norman Levitt, and Martin W. Lewis (eds.) (1996). *The Flight from Science and Reason.* New York: New York Academy of Sciences.

Gupta, R. C. "Spread and Triumph of Indian Numerals." *Indian Journal of History of Science* (1983) 13: 23–38.

Habib, S. Irfan and Dhruv Raina (eds.) (1999). *Situating the History of Science: Dialogues with Joseph Needham.* Delhi: Oxford University Press.

Hall, Rupert (1954). *The Scientific Revolution, 1500–1800: The Formation of the Modern Scientific Attitude.* London: Longmans.

———— (1962). "General Introduction" to Marie Boas Hall's *The Scientific Renaissance: 1450-1630.* London: Collins.

Hanley, Terry. "St Thomas' Use of al-Ghazali's *Maqasid al-Falsifa.*" *Mediaeval Studies* (1982) 44: 243–270.

Hanson, Norwood Russell (1961). *Patterns of Discovery: An Inquiry into the Conceptual Foundations of Science.* Cambridge: Cambridge University Press.

Harding, Sandra (1998). *Is Science Multicultural? Postcolonialisms, Feminisms and Epistemologies.* Bloomington: Indiana University Press.

Harris, Steven J. "Confession-Building, Long-Distance Networks, and the Organization of Jesuit Science." *Early Science and Medicine* (1996) 1: 287–318.

Harvey, William (1963). *The Circulation of the Blood, and Other Writings.* Trans. Kenneth J. Franklin. London: Dent.

Henderson, John (1984). *The Development and Decline of Chinese Cosmology.* New York: Columbia University Press.

Henry, John (2002). *The Scientific Revolution and the Origins of Modern Science.* New York: Palgrave.

Herodotus (1998). *The Histories.* Trans and Intr. Robin Waterfield and Carolyn Dewald. Oxford: Oxford University Press.

Hess, David J. (1995). *Science and Technology in a Multicultural World: The Cultural Politics of Facts and Artefacts.* New York: Columbia University Press.

Hetherington, Norriss S. "Plato and Eudoxus: Instrumentalists, Realists, or Prisoners of Themata?" *Studies in History and Philosophy of Science* (1996) 27(2): 271–289.

Ho, Peng Yoke (2003). *Chinese Mathematical Astrology: Reaching out to the Stars.* London: Routledge.

Hobson, John M. (2004). *The Eastern Origins of Western Civilisation.* Cambridge: Cambridge University Press.

Hogendijk, J. P. (2003). *The Enterprise of Science in Islam: New Perspectives.* Cambridge, MA: MIT Press.

Holton, Gerald (1973). *Thematic Origins of Scientific Thought: Kepler to Einstein.* Cambridge, MA: Harvard University Press.

———— (1993). *Science and Anti-Science.* Cambridge, MA: Harvard University Press.

Hoodbhoy, Pervez (1991). *Islam and Science: Religious Orthodoxy and the Battle for Rationality.* London and New Jersey: Zed Books.

Hooykaas, R. (1973). *Religion and the Rise of Modern Science.* Edinburgh: Scottish Academic Press.

Huff, Toby E. (2003). *The Rise of Early Modern Science: Islam, China, and the West.* Cambridge: Cambridge University Press.

Hugonnard-Roche, H. (1996). "The Influence of Arabic Astronomy in the Medieval West." In vol. 1, Rashed and Morelon (1996), pp. 284–305.

Huntington, Samuel P. (1996). *The Clash of Civilizations and the Remaking of World Order.* New York: Simon & Schuster.

Ifrah, Georges (2000). *The Universal History of Numbers: From Prehistory to the Invention of the Computer.* New York: John Wiley & Sons.

Iqbal, Mohammad (1934). *The Reconstruction of Religious Thought in Islam.* London: Oxford University Press.

Iqbal, Muzaffar (2002). *Islam and Science.* Burlington, VT: Ashgate.

Jay, Martin (1993). *Downcast Eyes: The Denigration of Vision in Twentieth-Century French Thought.* Berkeley: University of California Press.

Jones, W. T. (1969). *The Classical Mind: A History of Western Philosophy.* New York: Harcourt Brace & World.

Joseph, George Gheverghese (2000). *The Crest of the Peacock: Non-European Roots of Mathematics.* Princeton and Oxford: Princeton University Press. [First published by I.B. Tauris, 1991.]

Joy, Lynn Sumida (1987). *Gassendi the Atomist: Advocate of History in an Age of Science.* Cambridge: Cambridge University Press.

Kant, Immanuel (1950). *Critique of Pure Reason.* Trans. Norman Kemp Smith. New York: Macmillan.

Kargon, Robert Hugh (1966). *Atomism in England from Hariot to Newton.* London: Oxford University Press.

Kennedy, Edward Stewart (1998). *Astronomy and Astrology in the Medieval Islamic World.* Aldershot: Ashgate.

Keynes, J. Maynard (1947). *Newton Tercentenary Celebrations.* Cambridge: Cambridge University Press, pp. 27–34.

King, David (1993). *Islamic Mathematical Astronomy.* Aldershot: Variorum.

———— (2000) "Mathematical Astronomy in Islamic Civilisation." In Selin (2000), pp. 585–613.

Koertge, Noretta (ed.) (2000). *A House Built on Sand: Exposing Postmodern Myths about Science.* Oxford: Oxford University Press.

Koestler, Arthur (1964). *The Sleepwalkers: A History of Man's Changing Vision of the Universe.* London: Penguin Books.

Koyré, Alexandre (1957). *From the Closed World to the Infinite Universe.* Baltimore and London: Johns Hopkins University Press.

Kuhn, Thomas (1957). *The Copernican Revolution: Planetary Astronomy in the Development of Western Thought.* Cambridge, MA: Harvard University Press.

——— (1970). *The Structure of Scientific Revolutions.* Chicago: Chicago University Press.

Kumar, Deepak (ed.) (1991). *Science and Empire: Essays in Indian Context (1700–1947).* Delhi: Anamika Prakashan.

Kurtz, Paul and Timothy Madigan (eds.) (1994). *Challenges to the Enlightenment: In Defence of Reason and Science.* New York: Prometheus Books.

Landes, David S. (2000). *Revolution in Time: Clocks and the Making of the Modern World.* Cambridge, MA: Harvard University Press.

Lattis, James M. (1995). *Between Copernicus and Galileo: Christoph Clavius and the Collapse of Ptolemaic Cosmology.* Chicago: University of Chicago Press.

Leaman, O. "Averroes and the West." In Wahba and Abousenna (1996), pp. 53–67.

Lindberg, David (1992). *The Beginnings of Western Science: The European Scientific Tradition in Philosophical, Religious, and Institutional Context, 600 BC to AD 1450.* Chicago: University of Chicago Press.

Linden, Stanton J. (ed.) (2003). *The Alchemy Reader: From Hermes Trismegistus to Isaac Newton.* Cambridge: Cambridge University Press.

Lloyd, Geoffrey (2004). *Ancient Worlds, Modern Reflections: Philosophical Perspectives on Greek and Chinese Science and Culture.* New York: Oxford University Press.

Lloyd, Geoffrey and Nathan Sivin (2002). *The Way and the Word: Science and Medicine in Early China and Greece.* New Haven: Yale University Press.

Lu Gwei-Djen and Joseph Needham (1980). *Celestial Lancets: A History and Rationale of Acupuncture and Moxa.* Cambridge: Cambridge University Press.

Luthy, Christoph, John Murdoch, and William Newman (eds.) (2001). *Late Medieval and Early Modern Corpuscular Matter Theories.* Leiden: Brill.

Maimonides, Moses (1956). *The Guide for the Perplexed.* New York: Dover Publications.

Mallayya, Madhukar V. "Geometric Approach to Arithmetical Progressions from Nilakantha's *Aryabhatiyabhasya* and Sankara's *Kriyakramakari*." In Aryabhateeyam Proceedings (2002) pp. 143–147.

Margolis, Howard (2002). *It Started With Copernicus: How Turning the World Inside Out Led to the Scientific Revolution.* New York: McGraw-Hill.

Marmura, Michael (1975). "Ghazali's Attitude to the Secular Sciences and Logic." In *Essays on Islamic Philosophy and Science.* Ed. G. Hourani. Albany: SUNY Press, pp. 185–215.

McAleer, Graham. "Who Were the Averroists of the Thirteenth Century?: A Study of Siger of Brabant and Neo-Augustinians in Respect of the Plurality Controversy." *Modern Schoolman* (1999) 76(4): 273–292.

McClellan, James E., III, and Harold Dorn (1998). *Science and Technology in World History: An Introduction.* Baltimore: Johns Hopkins University Press.

McEvilley, Thomas (2002). *The Shape of Ancient Thought: Comparative Studies in Greek and Indian Philosophies*. New York: Allworth Press.

McGreal, Ian (ed.) (1995). *Great Thinkers of the Eastern World: The Major Thinkers and the Philosophical and Religious Classics of China, India, Japan, Korea, and the World of Islam*. New York: Harper Collins Publishers.

Mendelsohn, Everett (ed.) (1984). *Transformation and Tradition in the Sciences*. Cambridge: Cambridge University Press.

Merchant, Carolyn (1980). *The Death of Nature: Women, Ecology, and the Scientific Revolution*. New York: Harper & Row.

Merton, Robert K. (1970). *Science, Technology and Society in Seventeenth-Century England*. New York: Harper & Row.

Moody, E. A. "Galileo and Avempace: The Dynamics of the Leaning Tower Experiment." *Journal for the History of Ideas* (1951) 12: 163–193 and 375–422.

Mungello, David E. A. "On the Significance of the Question 'Did China Have Science?'" *Philosophy East and West* (1972) 22: 467–478.

Nagasaka, Francis (1998). *Japanese Studies in the Philosophy of Science*. Dordrecht: Kluwer Academic.

Nakamura, Hajime. (1975). *Parallel Developments: A Comparative History of Ideas*. London: Routledge & Kegan Paul.

Nakayama, S., and Sivin N. (1973). *Chinese Science: Explorations of an Ancient Tradition*. Cambridge, MA: MIT Press.

Nanda, Meera (2003). *Prophets Facing Backward: Postmodern Critiques of Science and Hindu Nationalism in India*. New Brunswick, NJ: Rutgers University Press.

Nandy, Ashis (ed.) (1988). *Science, Hegemony and Violence: A Requiem for Modernity*. Delhi: Oxford University Press.

Narayan, Uma (2000). *Decentering the Center: Philosophy for a Multicultural, Postcolonial, and Feminist World*. Bloomington, Ind: Indiana University Press.

Nasr, Seyyed Hossein (1968). *Science and Civilization in Islam*. Cambridge, MA: Harvard University Press.

——— (1976). *Islamic Science: An Illustrated Study*. Kent, UK: World of Islam Festival Publishing Co.

——— "Islam." In Sharma (ed.) (1993).

Needham, Joseph (1954, 1956). *Science and Civilization in China*. Vol. 1 and Vol. 2. Cambridge: Cambridge University Press.

——— (1958). *Chinese Astronomy and the Jesuit Mission: An Encounter of Cultures*. London: The China Society.

——— (1969). *The Grand Titration: Science and Society East and West*. London: Allen & Unwin.

——— (1970). *Clerks and Craftsmen in China and the West*. Cambridge: Cambridge University Press.

——— (1979). *Within the Four Seas: The Dialogue of East and West*. Toronto: University of Toronto Press.

Nelson, Benjamin. "Sciences and Civilizations, 'East' and 'West': Joseph Needham and Max Weber." In Seeger and Cohen (1974), pp. 445–493.

Neugebauer, Otto (1969). *The Exact Sciences in Antiquity*. New York: Dover.

Newton, Isaac (1730). *Opticks*. Reprint, New York: Dover, 1952.

O'Malley, John, Gauvin Alexander, and Stephen Harris (eds.) (2005). *The Jesuits: Cultures, Sciences, and the Arts, 1540–1773*. Toronto: University of Toronto Press.

Omar, Saleh Beshara (1997). *Ibn al-Haytham's Optics: A Study of the Origins of Experimental Science*. Minneapolis, MN: Bibliotheca Islamica.

Osler, Margaret (1994). *Divine Will and the Mechanical Philosophy: Gassendi and Descartes on Contingency and Necessity in the Created World*. Cambridge: Cambridge University. Press.

——— (ed.) (2000). *Rethinking the Scientific Revolution*. Cambridge: Cambridge University Press.

Pacy, Arnold (1990). *Technology in World Civilization: A Thousand Year History*. Cambridge, MA: MIT Press.

Park, David (1997). *The Fire within the Eye: A Historical Essay on the Nature and Meaning of Light*. Princeton, NJ: Princeton University Press.

Pickstone, John V. (2000). *Ways of Knowing: A New History of Science, Technology, and Medicine*. Manchester: Manchester University Press.

Pingree, David. "The Recovery of Early Greek Astronomy from India." *Journal for the History of Astronomy* (1976) 7: 109–123.

Polo, Marco (1958). *The Travels of Marco Polo*. Trans. Ronald Latham. Harmondsworth, England: Penguin Classics.

Pomeranz, Kenneth (2000). *The Great Divergence*. Princeton, NJ: Princeton University Press.

Porter, Roy, and Mikulas Teich (1992). *The Scientific Revolution in National Context*. Cambridge: Cambridge University Press.

Qadir, Chaudhry Abdul (1988). *Philosophy and Science in the Islamic World*. London: Croom Helm.

Qian, Wen-Yuan (1985). *The Great Inertia: Scientific Stagnation in Traditional China*. London: Croom Helm.

Rahman, Abdur (ed.) (2000). *History of Indian Science, Technology, and Culture, A.D. 1000–1800*. Delhi: Oxford University Press.

Rajagopal, C. T., and M. S. Rangachari. "On an Untapped Source of Medieval Keralese Mathematics." *Archive for History of Exact Sciences* (1978) 18: 89–101.

Raju, C. K. "Computers, Mathematics Education, and the Alternative Epistemology of the Calculus in the *Yuktibhasa*." *Philosophy East and West* (2001) 51(3): 325–361.

Ramakrishnan, P. "Some Aspects of Development in Kerala Mathematics." In *Aryabhateeyam Proceedings* (2002) pp. 137–141.

Rapson, Helen (1982). *The Circulation of Blood: A History*. London: Frederick Muller.

Rashed, Roshdi and Régis Morelon (eds.) (1996). *Encyclopaedia of the History of Arabic Science*, 3 vols. London: Routledge.

Rawlinson, H. G. "Early Contacts Between India and Europe." In Basham (ed.) (1975).

Rochberg, Francesca. "A Consideration of Babylonian Astronomy within the Historiography of Science." *Studies in History and Philosophy of Science* (2002) 33A(4): 661–684.

Ronan, Colin (1978, 1981). *The Shorter Science and Civilization in China: An Abridgement of Joseph Needham's Original Text.* Vol. 1 and Vol. 2. Cambridge: Cambridge University Press.

———— (1983). *The Cambridge Illustrated History of the World's Science.* Cambridge: Cambridge University Press.

Rorty, Richard (1980). *Philosophy and the Mirror of Nature.* Princeton, NJ: Princeton University Press.

Rosinska, Grazyna. "Nasir al-Din al-Tusi and Ibn al-Shatir in Cracow?" *Isis* (1974) 65: 239–243.

Ross, Andrew (ed.) (1996). *Science Wars.* Durham: Duke University Press.

Sabra, A. I. (1988). "Science, Islamic." In *Dictionary of the Middle Ages.* Vol. 2. Ed. J. R. Strayer. New York: Scribner's, pp. 81–89.

———— "The Andalusian Revolt against Ptolemaic Astronomy: Averroes and al-Bitruji." In Mendelsohn (1984), pp. 133–153.

Said, Edward (1994). *Culture and Imperialism.* London: Vintage.

Saliba, George (1994). *A History of Arabic Astronomy: Planetary Theories during the Golden Age of Islam.* New York and London: New York University Press.

———— "Arabic Planetary Theories after the Eleventh Century AD." In Rashed and Morelon (1996), vol. I, pp. 58–127.

Sangwan, Satpal. "Why Did the Scientific Revolution Not Take Place in India?" In Kumar (1991), pp. 31–40.

Sardar, Ziauddin (ed.) (1988). *The Revenge of Athena: Science, Exploitation, and the Third World.* London: Mansell.

———— "Above, Beyond and at the Center of the Science Wars: A Postcolonial Reading." In K. M. Ashman and P. S. Baringer (eds.) (2001).

Sarma, K. V. (1972). *A History of the Kerala School of Hindu Astronomy.* Hoshiarpur, India: Vishveshvaranand Institute.

Sarma, K. V. and S. Hariharan. "*Yuktibhasa* of Jyesthadeva: A Book of Rationales in Indian Mathematics and Astronomy—An Analytical Appraisal." *Indian Journal of History of Science* (1991) 26(2): 185–207.

Sarton, George (1975). *Introduction to the History of Science.* Vol. 1. New York: Krieger.

Saunders, J. J. "The Problem of Islamic Decadence." *Journal of World History* (1963) 7: 701–720.

Schafer, Edward (1977). *Pacing the Void: T'ang Approaches to the Stars.* Berkeley: University of California Press.

Seeger, Raymond J. and Robert S. Cohen (eds.) (1974). *Philosophical Foundations of Science.* Boston Studies in the Philosophy of Science Series. Vol. II. Dordrecht: Reidel.

Selin, Helaine (ed.) (1997). *Encyclopaedia of the History of Science, Technology and Medicine in Non-Western Cultures.* Dordrecht: Kluwer Academic Publishers.

———— (2000). *Astronomy Across Cultures: The History of Non-Western Astronomy.* New York: Springer.

Shankman, Steven and Stephen Durrant (eds.) (2002). *Early China/Ancient Greece: Thinking through Comparisons.* New York: SUNY Press.

Shapin, Steven (1996). *The Scientific Revolution*. Chicago: University of Chicago Press.

Sharma, Arvind (1993). *Our Religions*. New York: Harper Collins.

Sharrock, Wes, and Rupert Read (2002). *Kuhn: Philosopher of Scientific Revolutions*. Malden, MA: Polity Press.

Shen, Vincent and Tran Van Doan (eds.) (1995). *Philosophy of Science and Education: Chinese and European Views*. Washington: Council for Research in Values and Philosophy.

Shinki, Ichikawa. "Will We Ever Know What the Chinese Knew?" *Nature* (2000) 406 (6797): 673.

Shiva, Vandana (1988). *Staying Alive: Women, Ecology and Development*. London: Zed.

———— (1997). *Biopiracy: The Plunder of Nature and Knowledge*. Boston, MA: South End Press.

Sivin, Nathan. "Why the Scientific Revolution Did Not Take Place in China—or Didn't It?" In Mendelsohn (1984), pp. 531–554.

———— (ed.) (1995). *Science in Ancient China: Researches and Reflections*. Aldershot: Variorum.

Smart, Ninian (1999). *World Philosophies*. New York: Routledge.

Sriram, M. S. (2002). "Model of Planetary Model in the Kerala School of Astronomy." In *Aryabhateeyam Proceedings* (2002) pp. 105–113.

Sun, Xiaochun. "Crossing the Boundaries Between Heaven and Man: Astronomers in Ancient China." In Selin (2000), pp. 423–454.

Swerdlow, N. M., and O. Neugebauer (1984). *Mathematical Astronomy in Copernicus's "De Revolutionibus."* New York: Springer Verlag.

Tachau, Katherine (1988). *Vision and Certitude in the Age of Ockham: Optics, Epistemology, and the Foundations of Semantics, 1250–1345*. Leiden: Brill.

Teresi, Dick (2002). *Lost Discoveries: The Ancient Roots of Modern Science from the Babylonians to the Maya*. New York and London: Simon & Schuster.

Turner, Howard (1997). *Science in Medieval Islam: An Illustrated Introduction*. Austin: University of Texas Press.

Udías, Agustín. "Jesuit Astronomers in Beijing, 1601–1805." *Quarterly Journal of the Royal Astronomical Society* (1994) 35: 463–478.

Vattimo, Gianni (1988). *The End of Modernity: Nihilism and Hermeneutics in Postmodern Culture*. Trans. with introduction by Jon R. Snyder. Baltimore: Johns Hopkins University Press.

Veith, Ilza (trans.) (1966). *Huang Ti Nei Ching Su Wen: The Yellow Emperor's Classic of Internal Medicine*. Berkeley: University of California Press.

Wahba, Mourad, and Mona Abousenna (eds.) (1996). *Averroes and the Enlightenment*. Amherst, NY: Prometheus Books.

Wallace, W. A. (1991). *Galileo, the Jesuits and the Medieval Aristotle*. Hampshire, UK: Varorium.

Webster, C. (1982). *From Paracelsus to Newton: Magic and the Makings of Modern Science*. Cambridge: Cambridge University Press.

Weinberg, Steven (2001). *Facing Up: Science and Its Cultural Adversaries*. Cambridge, MA: Harvard University Press.

Westfall, Richard (1977). *The Construction of Modern Science: Mechanisms and Mechanics.* Cambridge: Cambridge University Press.

Whewell, W. (1847). *The Philosophy of the Inductive Sciences.* London: Parker.

White, L., Jr. (1962). *Medieval Technology and Social Change.* Oxford: Oxford University Press.

———— "Contributions to a Review Symposium on Science and Civilization in China." *Isis* (1984) 75(276): 172–179.

Wiebe, Donald (1991). *The Irony of Theology and the Nature of Religious Thought.* Montreal: McGill-Queen's University Press.

Wolf, Eric (1982). *Europe and the People without History.* Berkeley: University of California Press.

Wolfson, Harry Austryn (1976). *The Philosophy of the Kalam.* Cambridge, MA: Harvard University Press.

Wolpert, Lewis (1993). *The Unnatural Nature of Science.* Cambridge, MA: Harvard University Press.

Yates, F. A. (1969). *Giordano Bruno and the Hermetic Tradition.* Chicago: University of Chicago Press.

Name Index

Subject Index